The Sea and International Relations

MANCHESTER
1824

Manchester University Press

The Sea and International Relations

Edited by

Benjamin de Carvalho and Halvard Leira

MANCHESTER UNIVERSITY PRESS

Published by Manchester University Press
Oxford Road, Manchester M13 9PL

www.manchesteruniversitypress.co.uk

British Library Cataloguing-in-Publication Data
A catalogue record for this book is available from the British Library

ISBN 978 1 5261 5510 8 hardback

First published 2022

Typeset by Newgen Publishing UK

To Torbjørn, Dick and Yale

Contents

Contributors

Filippa Sofia Braarud holds a Master of Arts in International Security from Sciences Po Paris and Moscow State Institute of International Relations, and a Bachelor of Arts in International Justice from Leiden University College The Hague. Her academic interests lie at the intersection between international law, energy policy, technology and security.

Benjamin de Carvalho is Research Professor at the Norwegian Institute of International Affairs (NUPI) in Oslo. He holds a PhD from the University of Cambridge, where he worked on the reformations and state formation. He is currently the Principal Investigator of the Empires, Privateering and the Sea Project (EMPRISE), funded by the Research Council of Norway, which this current volume is a part of. He has written extensively on early modern historical International Relations, and been active in the Historical International Relations section of the International Studies Association (ISA) since its inception and in a number of functions. His latest publications include *Status and the Rise of Brazil* (Palgrave Macmillan, co-edited with Maria Jumbert and Paulo Esteves, 2020). With Halvard Leira he has also co-edited the four-volume *Historical International Relations* (SAGE, 2015), and the recently published *Routledge Handbook of Historical International Relations* (with Halvard Leira and Julia Costa López, 2021).

Alejandro Colás is Professor of International Relations in the Politics Department at Birkbeck College, University of London. He is the co-author, with Liam Campling, of *Capitalism and the*

Sea: The Maritime Factor in the Making of the Modern World
(Verso, 2021) and co-author of *Food, Politics and Society: Social
Theory and the Modern Food System* (University of California
Press, 2018).

Julia Costa López is Senior Lecturer in History and Theory of
International Relations at the University of Groningen. Her
research interests lie at the intersection of International Relations
and history of international political thought, particularly in the
late-medieval and early modern periods. She is one of the editors
of the *Routledge Handbook of Historical International Relations*
and her work has been published in journals such as *International
Organization*, *Review of International Studies*, and *International
Studies Review*.

Andonea Jon Dickson is a doctoral candidate and teaching associate in the School of Politics and International Relations, Queen
Mary University of London. Andonea's research brings the maritime to the fore in the analysis of migration regulation in the
Mediterranean. By considering the human and non-human elements in the regulation of human migration, her work emphasises
the entanglements and mobilities which lead to migrants being
contained and excluded in maritime geographies.

Kerry Goettlich is lecturer in International Security at the University
of Reading. He previously completed his PhD in International
Relations at the London School of Economics, where he was an
editor of *Millennium: Journal of International Studies*. His current
project examines the historical emergence of scientific practices
underlying modern territoriality, such as border surveying, as they
emerged in seventeenth-century colonial North America and were
globalised in the late nineteenth century. His work has appeared in
the *European Journal of International Relations* and the *Oxford
Research Encyclopedia of International Studies*.

Xavier Guillaume teaches at the Rijksuniversiteit Groningen, the
Netherlands, in the Department of International Relations and
International Organisations.

Halvard Leira is Research Professor at the Norwegian Institute of International Affairs (NUPI). He has published extensively in English and Norwegian on international political thought, historiography, foreign policy and diplomacy, more often than not with an emphasis on historical International Relations. His work has appeared in journals including *International Studies Quarterly*, *Review of International Studies*, *Millennium*, *Leiden Journal of International Law* and *Cooperation and Conflict*. Leira is co-editor of *International Diplomacy* (SAGE, 2013), *Historical International Relations* (SAGE, 2015), *The Routledge Handbook of Historical International Relations* (Routledge, 2021) and the current volume. He is currently serving as Associate Editor of the *European Journal of International Relations*. He is former section chair and programme chair of the Historical International Relations section of the International Studies Association and was programme chair of the European International Studies Association's Pan European Conference in 2018.

Maria Mälksoo is Senior Researcher at the Centre for Military Studies at Copenhagen University. She is the author of *The Politics of Becoming European: A Study of Polish and Baltic Post-Cold War Security Imaginaries* (Routledge, 2010), and a co-author of *Remembering Katyn* (Polity, 2012). Her work on memory politics, ontological security, liminality and European security politics has appeared in various International Relations journals and edited volumes. Her present research focuses on rituals in world politics, particularly in relation to extended deterrence, and memory laws in Eastern Europe. She is currently editing the *Handbook on the Politics of Memory* (Edward Elgar, forthcoming).

Mark Shirk teaches International Relations at Bucknell University. He received his PhD in Government and Politics from the University of Maryland in 2014. His work deals with the role of transnational processes such as piracy, terrorism and ecological change in the formation of the state and global order. He found his sea legs dressing up as a pirate at university and sailing in the Delaware Bay. Indeed, most of his published work is about piracy. You can find said work in *International Studies Quarterly*, *International Studies Review*, and *Terrorism and Political Violence*, among others.

Jessica K. Simonds is a third-year PhD candidate in International Studies at the School of History, Anthropology, Philosophy, and Politics at Queen's University Belfast. She holds a BA in International Politics and Conflict Studies and an MA in Violence, Terrorism, and Security, also from Queen's University. Her research interests include counter-piracy practice, critical security studies, maritime insurance and seafarer welfare.

Preface and acknowledgements

We would have loved to start this volume by professing our lifelong passion for all things maritime, recounting childhood summers on sailboats criss-crossing the fjords and archipelagos of Norway. But doing so would be to engage in falsehood. Although Halvard grew up close to water and did his fair share of swimming and fishing, he spent more time in the mountains than in the fjords. And although Benjamin once built a 22-foot-long Viking ship (modelled after the Norwegian *Nordlandsbåt*), his upbringing in Geneva restricted his maritime aspirations to the Swiss mountains and his adventures on water to the more riberian Lac Léman. A true *marin d'eau douce*, as Captain Haddock would have said. Furthermore, although we have indeed read some sea novels, neither Horatio Hornblower nor Jack Aubrey were part of our literary pantheon. When we are 'pining for the fjords', like the Norwegian Blue parrot immortalised by Monty Python, the cause must be sought elsewhere.

What brought us to the sea was history. One of us spent years studying Tudor England, a time and place where the sea was thrust into focus. The other spent years studying diplomatic history, digging deep into consular affairs and thus ports and sailors. One thing led to the other, and through some serendipitous quirks we found ourselves in the late 2000s writing about consuls and privateers. Obscure research papers and conference presentations led us to focus even more on privateering, and subsequently to funding for further digging with the project EMPRISE (Empires, Privateers and the Sea), which the Research Council of Norway generously funded (Grant 262657) in 2016. As EMPRISE has stood for the financing of this volume, without this funding, there most likely would have been no book, so our most sincere gratitude is in order.

Our aim with EMPRISE is to highlight and break down the
dichotomies which structure too much of our thinking in
International Relations (states/empires, land/sea). While our
starting point for EMPRISE was historical, we soon decided that
a necessary step would be to engage the sea in its own right.
Realising that we could not do that alone, we invited submissions
to a workshop on 'Rethinking the Sea in IR: Space, Time, Politics
& Change' under the EWIS umbrella, in Krakow in 2019. EWIS
(European Workshops in International Studies) is a fantastic forum
for large-group thematic discussions, organised by the European
International Studies Association (EISA). We would like to
thank EISA and the EWIS organisers, Maria Mälksoo and Artur
Gruszczak, for the opportunity to come to Krakow. In Krakow, two
days of spirited discussions and joyful interaction among twenty-
five engaged scholars led us to conclude that it was both possible
and necessary to produce this edited volume. All of the chapters in
the present volume were first presented as drafts in Krakow, apart
from the conclusion, which was written by our eminent discussants,
Julia Costa López and Xavier Guillaume. They deserve thanks and
thanks again for their comments, for their intellectual engagement
with the project, for their conclusion and for being A+ colleagues
and friends. Alejandro Colás, author of the first chapter in this
volume, also deserves a special mention, as he co-edited the first
volume in which our writings on privateering were published.
Without that invitation to the workshop he and Bryan Mabee
organised in London in 2009, this book may never have been
published. All this to say that research careers and interests move
in mysterious ways, and we are extremely grateful to the people
we have encountered on our travels, who have not only provided
us with generous comments, nods and suggestions, but also have
become great friends in the process.

Some of those friends have found their way into this volume, and
we would like to thank all our authors profusely for their prompt
replies, their patience and their overall enthusiasm for the project,
even in relatively bleak times. We thought that a volume such as
this required a diverse group of researchers, and the end result has
proved us right. The chapters in the volume have been written by
colleagues from different places, and at different stages in their
career. This combination proved invaluable to our discussions and

to the quality of the final product. With some luck, the launch of this book will coincide with a loosening of COVID restrictions, and we will all be able to meet again over seafood and Sancerre.

As with all projects and books, authors and editors rack up debts. This is our opportunity to pay some of them off. At our institutional home, NUPI, our colleagues always deserve thanks for their support, even if they find some of our work quaint. Special thanks are due to Ole Jacob Sending, Guri Bang, Paul Beaumont, Elana Wilson Rowe, and Roman Vakulchuk for their comments on the introduction. Within the EMPRISE project, and long before it started, Jens Bartelson has been an invaluable friend, supporter and mentor. Also within EMPRISE, John Hobson has been a constant source of inspiration. Ten years after we made 'big bangs' with John, we hope this volume will make a few big splashes.

At MUP, we have had the luck to work with Jonathan De Peyer, who was, coincidentally also the editor for the above-mentioned volume edited by Colás and Mabee many years ago, Robert Byron and Lucy Burns. They have provided much-needed support and encouragement, and made the editing process a lot less stressful than it could have been.

The publication of such a volume is also the place to thank those we meet with for seafood (and other food) everyday, and whose generosity and understanding of time constraints, tight deadlines and time spent away for fieldwork, archives or conferences is the condition of possibility of our academic endeavours. Benjamin thanks Ingrid, and their daughters Saskia and Katja, while Halvard thanks Ina, and his children Håkon and Vilja. Without their patience and understanding, this volume most certainly would not have seen the light. As always, this work is for them. We hope to be taking them to the seaside for celebrations soon.

Finally, it is hard to overstate the importance of the support Torbjørn L. Knutsen, Yale Ferguson and Richard Mansbach have given to young scholars with an interest in historical International Relations. We cherish their friendship and admire their scholarship, and without them as pathbreakers, there probably would have been no project (and thus, no book). It is with gratitude and pride that we dedicate this volume to them.

Introduction: staring at the sea

Benjamin de Carvalho and Halvard Leira

'What would we do without the oceans? Carry the boats?' This joke from coastal Norway is not only telling about the extent to which maritime life for many people is an inherent part of everyday life, but can also be read as a biting commentary about how the academic discipline of International Relations (IR) has to a large degree been content with carrying the boats, pretending as if the oceans were not there, as if their existence did not matter to our analyses of international relations and as if the boats made perfect sense on dry land.

For an academic discipline which boasts of dealing with the global, it is peculiar that IR has stubbornly, repeatedly and obsessively limited its gaze to a little less than thirty per cent of the globe – the landed part. Save a few exceptions, in region-specific and topic-specific parts of the discipline, IR has been a pathological 'landlubber discipline': either refusing to deal with the sea or treating it as land. This is an unfortunate state of affairs. With sea-level rise, depletion of fish-stocks, plastic pollution and piracy making the news repeatedly and constantly, it is obvious that the sea matters in international relations. It should also matter to the discipline studying these relations. In related disciplines (see e.g. Paine, 2013 for History) burgeoning literatures have recast the importance of the sea for understanding not only the past, but also key parts of our current predicament. Time has come for IR to catch up; to launch our boats into the sea. This would benefit the discipline, but it would also make contributions to a better understanding of the sea. With its diverse approaches to conflict, cooperation, and political co-existence, IR has obvious insights to bring to the study of the sea.

By addressing the sea, IR could stress the *political dimension* of maritime orders often taken for granted. We stress this political dimension in particular as an antidote to the long-standing depoliticising tradition of seeing the sea as a global commons. While this approach has provided interesting insights and important ways of thinking about policy, it has served to elide how the notion of the sea as commons has rested on naval hegemony, and an active forgetting of the endemic oceanic violence of the centuries before the nineteenth century.

This volume breaks with the trend of oceanic amnesia in IR, with the goal of kickstarting a theoretical, conceptual and empirical conversation about the sea and International Relations. The authors address the sea head on through understanding what implications it holds for our analyses of global issues and international relations. They do so by focusing on key dimensions through which the sea has played a key role for global politics, categorised under three headings: (1) taming or mastering the sea (2) traversing the water and (3) controlling maritime resources (see further discussion of these dimensions, or tropes, below). On the one hand we seek to expand the horizons of IR by incorporating the sea, on the other hand we suggest what IR brings to the sea. Specifically, we believe that IR provides what one could call an *amphibious* approach. Rather than exchanging land with sea, we focus on the interplay between sea and land, and insist on the political character of the social space of the sea.

The aim of this volume is thus twofold; to bring the sea more explicitly into IR and to take IR to the sea. In this introduction we set the stage, discussing how and why IR has engaged (or not) with the sea, exploring what other disciplines can offer IR when staring at the sea and suggesting some possibilities for fruitful engagement. We first set the scene by discussing why the sea has been missing from IR and the challenges facing us when trying to theorise the sea. Then we engage with the developing literature in other disciplines from the last two decades, illustrating why an IR-take makes sense, and where there is room to expand on the existing IR-literature. The third section puts the focus on politics, circulation and control, before the last section lays out how the different chapters engage with these overarching topics.

Landfilling the globe, flattening the waves

Before engaging with other disciplines, it makes sense to explore why the sea has been marginalised in IR. Our argument is not that the sea has never been addressed in IR, but rather that the attempts made at bringing the sea into the discipline have tended to remain on the margins; on the shores, so to speak. As a result, in spite of these efforts which we will discuss in more details below, the sea has remained outside of the mainstream of the discipline, and few have paid attention to it when theorising international relations.

One reason for the lack of attention paid to oceans could simply be that they are harder to control physically, and that it is hard to theorise about political interaction if there is no permanent control over any stable place in which to interact. We find this way of thinking historically deficient. Until the age of steam, long-distance travel at sea relied on wind and currents, and most shipping could be found along well-known 'lanes' at sea. Early overseas empires did stake claims to these lanes, as some sort of extension of their landed power (Benton, 2011). Until the final third of the nineteenth century and the emergence of steam shipping, it could thus make some sense to claim that some state(s) controlled the sea, while other states might attempt to circumvent or undermine that control. Perhaps, then, the rise of steamships was the key part of the great transformation of the nineteenth century which created the world of IR (Buzan and Lawson, 2015)? In the sense that a world without controllable shipping lanes is a world where IR can be thought without much concern for the sea? Steam power allowed IR to forget the sea as space, and to count it simply as time.

To the extent that territoriality is a key building block of IR, the steam-powered move away from controllable sea lanes is one plausible reason why IR has shied away from the sea. This explanation is, however, not sufficient. We would argue that the failure of IR to engage with the sea also stems from the relatively ahistorical roots of the discipline. To us, the lack of intellectual engagement with the sea in IR seems obviously related to how the discipline emerged and has developed in a period of naturalised oceanic hegemony. In the centuries before the nineteenth, the sea figured prominently in much thinking about states, empires and the relations between them.

Hugo Grotius' writings are a case in point, based as they were to a large extent on how the emerging Dutch empire could and should relate to other polities at sea. During two centuries of established naval hegemony, the sea could be taken for granted in much theorising about the globe.

The difference between IR and International Law (IL) is instructive in this respect. IL indeed traces its roots back to the thinkers of Grotius' time, to topics not necessarily tied to territoriality and to questions concerning how to handle the lack of overarching authority at sea. Several of the key topics in the development of IL, such as freedom of navigation, distinctions between piracy and privateering and the rights of neutral shipping were directly tied to the sea, and even under the condition of naval hegemony, IL has continued to focus on the developing regulation of the ocean. This may indeed help us understand why the sea seems to be a more intrinsic part of IL than IR, as witnessed by the discussion in Chapter 7.

When IR emerged gradually from around 1900, questions of overseas empire were central, but more in the guise of imperial administration than with any relation to the sea. On the other hand, the emerging geopolitics of the same period had a strong focus on the ocean (see Ashworth, 2011 on Mackinder), a focus which remained until at least mid-century (Rosenboim, 2017), but more so in the subfield of strategic studies than in IR more generally. More often than not, if mentioned at all, the sea was simply there, as a space to traverse, or a place for resource-extraction. Of course, there are no rules without partial exception, which holds true in this case too. In their world-systems approach to sea power, Modelski and Thompson (1998: 5) emphasised that 'the modern world system is, characteristically and importantly, an oceanic system. ... The advent of the modern world system was at the same time also the onset of use and control of the seas on a global scale, hence the opening of an entirely new age of sea power'. Even so, the sea in this account remains a constant, a medium upon and beneath which, and from which, sea power can be projected.

The general forgetting of the historical importance of the oceans was perpetuated by the strong Central-European roots of much post-World War II IR theorising. If the experiences of continental Europe shaped IR theorising, it is hardly surprising that the oceans became a secondary concern. Yet, this omission is deeply problematic. Any historically grounded theorisation of IR which does not include consideration of oceans must be somewhat deficient. Oceanic travel was what connected the word into one global system, and from the rise of the Iberian empires to the fall of the British empire, all of the aspiring hegemons have relied on the capacity to deliver goods from overseas possessions to the imperial centre.

It is obviously possible to theorise international relations without acknowledging the sea. Most theories of IR are theories of what happens when landed entities engage with one another, completely ignoring the maritime domain. The sea is still more often than not merely an adjunct, a conveyor-belt or an obstacle to be overcome in the interaction between landed units. In what can be read as a *coda* to the geopolitical thinkers of the early twentieth century, John J. Mearsheimer (2001) for instance did refer prominently to the 'stopping power of water', and made much of the notion of the 'offshore balancer.' Yet, while it plays a key role in the theory, the role of water remains almost magical.[1]

The 'landedness' of the sea is prevalent in many studies of naval and sea power. As Geoffrey Till has shown, studies of sea power have remained largely within the trail of Mahan's famous dictum about the relative economy of naval power and its tremendous benefits in terms of global power projection: '[c]ontrol of the sea by maritime commerce and naval supremacy means predominant influence in the world [and] is the chief among the merely material elements in the power and prosperity of nations' (cited in Till, 2018: 1). The importance of the sea is of course central to any treatment of sea power, as Till makes clear, not the least because of key attributes of the sea. It is (1) a resource in and of itself, (2) a medium of transportation and exchange, (3) a medium of spreading information and the spread of ideas, and (4) can be understood as dominium (Till, 2018: 17). Beyond the 'transformative' power of the sea in that it has allowed less powerful states such as the UK dominion on a global scale, the sea is not taken as a starting point for theorising concepts of phenomena beyond the shore. The sea, then, remains largely an extension of land, and sea power is limited to being a corollary of land-based military power and a space which must be traversed in order for states to project their power beyond their narrow shores. The work of Andrew Lambert on sea power states follows those lines. To Lambert, the sea represented an opportunity for smaller states to develop inclusive, dynamic, outward-looking and progressive polities and cultures with the sea as their chief commercial and diplomatic resource (2018). All told, a considerable number of studies of maritime warfare and navies have been undertaken within the discipline. But where one could have expected these to highlight how maritime security was different from traditional (grounded) security, these have seen the maritime domain as an unquestioned extension of the terrestrial one.

As the examples above illustrate, it is not as if the sea has had no place in IR. There are specialised literatures on e.g. maritime security, fisheries regimes, the law of the sea, sea power, climate change and the sea, ocean governance and naval strategy. Likewise, there are regional literatures about specific parts of the ocean considering these and other topics. To return to the theme from the introduction, these literatures still come close to boat-carrying. They recognise the sea as an important organising feature of human life, but it is still just there, as space, time or resource. It does not matter in and of itself, it is not theorised or even part of any theoretical framework. We accept that oceans are notoriously hard to theorise as a constituent part of international relations, but that should not stop us from trying.

IR has not been alone in finding the ocean hard to think with. Until fairly recently, the sea has been elusive to most social scientists, and therefore also largely absent from their works. As Steinberg (2014: xv) noted, 'Since the sea is a space that cannot be located and that cannot be purely experienced, thalassography – sea-writing – presents a challenge [and] [i]t is no wonder that the social science literature on the sea as a holistic space of interspecies intersubjectivity is exceptionally sparse.' Steinberg's point here echoes the long-standing difficulty in grasping the specificities of the sea, as Mahan noted more than a century ago: 'Historians generally have been unfamiliar with the conditions of the sea, having as to it neither special interest nor special knowledge; and the profound determining influence of maritime strength upon great issues has consequently been overlooked (Mahan, 1890 [2018]:1).

Furthermore, the academic disciplines which have taken it upon themselves to master the study of space have been defined as 'earth writing' (geography) (Barnes and Duncan, 1992: 1), and have remained true to their etymological roots (Steinberg, 1999a, Peters, 2010; Anderson and Peters, 2014: 3). The maritime has tended to be relegated to either 'the backdrop to the stage on which the real action is seen to take place – that is, the land – or they are portrayed simply as the means of connection between activities taking place at coasts and in their interiors' (Mack, 2011: 19).

The ways in which the sea has figured in the modern imagination has done little to make its role more prominent in theorising the globe. At the most basic level, the sea is different from land. The sea is the other in the traditional binary which separates land and sea (Westerdahl, 2005: 13). As a consequence of this, 'The ocean can then be categorized as a space of nature to be fetishized, a space of alterity to be romanticized, or even a space beyond society to be forgotten. In each of these formulations, the ocean is classified as an object, a space of difference with a distinguishing ontological unity, the "other" in a land-ocean binary' (Steinberg, 2014: xiii). The consequences of this were pointed out decades ago, scholars 'bound by a European terrestrial bias, have accepted as *natural* the dominance of the land in understanding human interactions and relationships with environments' (Jackson, 1995: 87–8; emphasis original). This spatial othering has also been coupled with a material othering: 'This "naturalized" position of the oceans as marginal to the land, is, moreover, enforced through the liquid materiality of water. The sea's physical constitution renders it as intrinsically "other"; it is a fluid world rather than a solid one. Our normative experiences of the world centre on engagements on solid ground; rather than in liquid sea' (Anderson and Peters, 2014: 5).

The relative difficulty of accessing the sea has rendered it marginal and often addressed obliquely: 'in many ways the ocean seems to be a space more suitable for the literary essay or poem that reproduces difference even as it interrogates its foundations, for the policy analysis or military strategy that analyzes one particular ocean use while ignoring others, or for the philosophical tome that reduces the sea to a metaphor for flux and flow while ignoring the actual mobilities that are experienced by those who traverse or gaze upon its surface' (Steinberg, 2014: xv). Broadly speaking, as noted above, we could speak of three tropes which cover the way the sea has been dealt with in the Western tradition: (a) the sea as a space to be *tamed*, (b) the sea as a space to be *traversed*, and (c) the sea as a space to be *controlled*. The most prominent view of the sea, according to Mack has been the first one, namely that of 'a quint-essential wilderness, a void without community other than that temporarily established on boats crewed by those with the shared experience of being tossed about on its surface' (Mack, 2011: 17). This dovetails with the view of oceans as obstacles which needed to be tamed and mastered in order to establish domination over vast distances (Anderson and Peters, 2014: 1; Law, 1986; Ogborn, 2002). Added to this first trope of indomitable wilderness is the sea as a 'non-developable space' or an 'empty transportation sur-face, beyond the space of social relations' (Steinberg, 2001: 113). According to this largely mercantilist trope, then, the sea is a mere surface to be traversed; an empty space of circulation. Here, oceans and seas are mere 'spatial fillers to be traversed for the capital gain of those on land' (Anderson and Peters, 2014: 1; Steinberg, 2001), the 'vast void' between states (Steinberg, 2001: 113). Finally, one could add a third trope; one in which the sea is no different from land, simply an extension of it. Dovetailing with security studies or maritime security, the sea is yet another space – if not territory – which must be secured and controlled. At any rate, whichever trope the sea has been imagined through, it has not figured prominently in either one of them in its own right. Agency is located outside of the domain of the sea, and the sea is only there to be passively (albeit with some resistance) acted upon.

As Langewiesche has pointed out, 'Since we live on land, and are usually beyond the sight of the sea, it is easy to forget that our world is an ocean world' (2004: 3). One could further hypothe-sise that along with technological developments in shipbuilding and fisheries, fewer people experience the sea firsthand today as opposed to earlier centuries. The sea has gone from a space that had to be reckoned with in the daily activities of many people to a space of leisure. A number of tourists in the course of their leisure encounter the sea, yet these encounters of the sea are of a com-pletely different character than the experience of earlier centuries. It is perhaps not surprising then, that the social sciences have until recently been characterised by a certain 'seablindness' (Bueger and Edmunds, 2017). We now turn to how the situation has changed.

Thinking about the sea

The sea has made an obvious return over the last decades, in the news and in academe: '[r]ediscovered as a crucial space of globalization ... oceans have swung insistently into view in recent years. And slowly but surely, scholarly attention has followed' (Wigen, 2014: 1). The scientific attention has come in a variety of forms, often interdisciplinary, grouped under headings such as 'blue cultural studies'/ 'blue humanities' (Mentz, 2009; Gillis, 2013) or 'critical ocean studies' (DeLoughrey, 2017; 2019). Under these umbrella-terms, a wide range of topics and approaches can be found. Some use a maritime focus to reinterpret the past and present, while others suggest that engaging with the sea could (and even should) lead to a thoroughgoing rethinking of basic ontologies, epistemologies and ethics (Winkiel, 2019). These are important literatures, and both we and the other authors in this volume engage with them. Among other things, these literatures urge us to rethink the Eurocentric and gendered nature of many of the traditional approaches to the sea. Put bluntly, there are many other ways of approaching the oceans than the Western, masculine one. We deal with this explicitly in Chapter 3. However, for the purpose of bringing general IR and the sea closer together, we believe that the explorations must start closer to 'home', in neighbouring disciplines and how they can combine with IR.

Different landed disciplines face different kinds of questions when moving to sea. Geographers have suggested a focus on 'the sea as a holistic space of interspecies intersubjectivity ... characterized by the co-construction of maritime subjects – from sailors and swimmers to reefs and water molecules' (Steinberg, 2014: xv).[2] The corollary of this focus is the move we note above, from geography to 'thalassography'. From our perspective, this take is both too wide and too narrow. Too wide, in that we are not trying to understand the sea *as such* to the depth (pun intended) of the geographers. Too narrow in the sense that we are not writing the International Relations *of the Sea* but giving perspectives on how the sea matters *to* International Relations. Our vantage-point from International Relations also implies that we are not concerning ourselves with all inter-polity relations of, on and about the sea, but restrict our exploration to the last 600 years or so. Here we follow a diverse set of thinkers stressing how the age of exploration and mercantilism implied a spatial reconfiguration of land, sea and globe (Schmitt, [1942] 2015; Modelski and Thompson, 1998; Steinberg, 2001). In that sense, although we disagree with both the politics and the conclusion drawn by Schmitt ([1942] 2015),[3] we concur with the overarching notion that International Relations need to consider Land *and* Sea.

Much like geography and IR, most macro-historical accounts of political transformation have been 'landed' (Steinberg, 2001; see also the accounts in de Carvalho, Costa López and Leira, 2021) to the point where few theoretical statements take the sea into account, and sea power figures either as land power, its extension, or, as operating along the same logic (see Tilly, 1992: 94; see the critique in de Carvalho and Leira, 2021). For the most part, the sea is ignored or obliterated as a transport leg, or a neutral conduit for European expansion (Blakemore, 2013); counted as time rather than analysed as space (Blakemore, 2013). Yet, due to a lack of landed infrastructure, sea lanes were central to both trade and warfare throughout the early modern period (Subrahmanyam, 1996), and amphibious power – such as privateering – came to be crucial in securing control over these spaces (Trim and Fissell, 2006). One early take on this can be found in the work of Jan Glete. While most of his writings concentrated on navies and naval history, Glete also linked the issue of warfare at sea to the formation of states in Europe, forcefully demonstrating in a Tillian fashion the extent to which wars made states; just not the wars Tilly had covered. On the contrary, Glete shows the extent to which arming a navy required so much more long-term investment and bureaucratic expertise than landed warfare, that if war was a decisive factor in the making of states, it was naval warfare rather than traditional landed battles which provided the main impetus behind such a large scale political transformation (see, e.g. Glete, 2000). Glete was at the forefront of what has become a more sustained turn towards the oceans in global history.

In a recent opening statement, summarising two decades of oceanic history, Sivasundaram, Bashford and Armitage (2018) celebrate the diversity in how historians have dealt with the naval, the maritime and the ocean, and put special emphasis on ecology, space and time. These topics are covered in the ensuing pages as well, and from an International Relations point of view, we will add an emphasis on the political quality of the sea. While the sea can at one and the same time have both material and cultural dimensions (as demonstrated in the chapters of this volume; see also Sivasundaram, Bashford and Armitage, 2018: 16) as long as human interaction with the ocean involves more than one polity, there will be a political dimension to the sea. As noted earlier, only by bringing this political dimension to the fore can we begin to comprehend the extent to which the order at sea we take for granted today rests on centuries of challenges to hegemonic rule, culminating in the Anglo-American order that has been in place for close to a century and a half.

We should stress that seeing the ocean as political does not simply imply that the ocean is a space where politics is played out, it implies seeing the ocean as something with no inherent meaning, as something which is politically contested as such. Thus, our view of the ocean as inherently political implies that we are primarily concerned with the social construction of the ocean (Steinberg, 2001), and how different disciplines and approaches have contributed to shape the ocean. Furthermore, an insistence on the political quality of the sea does not imply state-centrism. One of the important contributions of the new wave of ocean studies discussed above, has been to de-privilege the state in analyses of the sea. This is not to say that states can be ignored, but to reiterate that they are only one of many potential actors in out study of the sea.

We could summarise, with Wigen, that historians have tended to emphasise the sea as 'a highway for intercontinental exchange', social scientists have approached the sea as 'an arena for conflict' and humanists have preferred to 'probe the contours of the oceanic imaginary in film and fiction, map and metaphor' (Wigen, 2014: 1). IR, with its methodological plurality and ontological richness, we believe, opens up for weaving many of these aspects together. Furthermore, as much of the work on the sea has tended to focus on specific spaces enclosed by specific seas or on definite spaces connected by individual oceans – what Wigen calls 'a burgeoning but fragmented body of work, framed within individual basins' (Wigen, 2014: 2) – IR holds the promise of a more general understanding of the sea in human experience and a more global take on its importance.

Rethinking the (IR) world with the sea

As we have laboured above, the sea has been conspicuously absent from theorising in IR. This lack of theorising has been bemoaned recently, with scholars pointing out for instance that '[w]here discussion has taken place, it has largely been secondary or subsidiary to studies on specific issues such as piracy, counter-piracy or illegal fishing, or to wider debates on sea power or the law of the sea. Only rarely has maritime security capacity-building been addressed on its own terms' (Bueger et al., 2019: 2). As the same authors acknowledge, addressing the sea in IR needs to incorporate land: 'Maritime security is not simply about the sea. The challenges it presents are closely interlinked with issues of development and security on land, in terms of both cause and effect' (Bueger et al., 2019: 3).

Thus, while oceans are distinctive spaces (Steinberg, 2001; Shilliam, 2015), to IR it is first and foremost the extent to which these spaces interact with other spaces and how the space of the sea contributes to shape that interaction and constitute both actors and power in specific ways which is the topic of the present volume. Rather than a wholesale adoption of the sea into IR and consequent obliteration of land, we believe that the most promising ground for the discipline lies in *combining* the work pioneered by Steinberg and Peters (2015) on the ocean 'as a dynamic environment of flows and continual recomposition where, because there is no static background, "place" can be understood only in the context of mobility' with perspectives emphasising the amphibious or liminal nature of social phenomena (see, e.g., Trim and Fissel, 2006; Klein and Mackenthun, 2004). To this, we would add the political quality of the oceans, leaving us with a set of approaches which combine sea, land and politics.

Recent contributions in IR showcase some of the promise of such combinations. Andrew Phillips and Jason Sharman's take on heterogeneity in the Indian Ocean (2015), or Jeppe Mulich's (2015) work on the coastal or maritime 'space between empires' incorporate, as Alejandro Colás has argued, 'the diversity or plurality of polities and territorialities fostered in large measure by the particularities of the sea, into our explanations of international relations' (2019). Likewise, Luis Lobo-Guerrero has dealt with how the sea structures global governance (2011; 2012; Lobo-Guerrero and Stobbe, 2016).

Charlotte Epstein too, has shown how maritime discourses offer a platform to rethink our (terrestrial) conceptual arsenal (2008). More historically, Alejandro Colás (2016; 2019; Campling and Colás, 2018; Colás and Mabee, 2010) has sought to bring the sea into analyses of the development of capitalism. Finally, our own work has sought to explore duality through liminal institution of privateering and amphibious nature and practice of privateers (Leira and de Carvalho, 2011). Yet while these contributions have all approached the sea and dealt with 'blue IR' to a larger extent than others, none of them put the sea squarely at the centre of their study, and as such they all fall short of giving a comprehensive view of the challenges the sea represents for research in IR.

As alluded to above, we need to tread cautiously when emphasising the distinctness of the sea. Indeed, where some scholars have called for a 'blue turn' in IR we wish to caution here against such an enthusiasm. Not only because we tend to think that the whole business of turns has gone overboard in IR, and that it has been reduced to a marketing device rather than an accurate description of the state of disciplinary developments, but also because, following Peters and Steinberg (2019), such a take overlooks the extent to which theorising from the sea offers new perspectives on the friction between sea and land and uncovers the amphibious nature of many social phenomena.

We believe that rather than to adopt a 'wet ontology' wholesale or even rethinking IR from the sea, we believe the true promise of bringing the sea lies in rethinking IR *with the sea*.[4] While we need to break with the traditional assumption that sea power is a corollary of landed political power, we cannot discard their relation altogether. Nor can we accept the distinction made by students of naval power only. Instead, we need to focus on how they interact, for instance through 'maritime contact zones' (Klein and Mackenthun, 2004). For however important the sea is, be it for transportation, extraction or strategic purposes, we must keep in mind that sea and land are not separate spaces but intertwined; seapower cannot do without a basis on land.

While there is a burgeoning scholarly literature on the sea, we agree with Kären Wigen that there is still a need for a better conceptual apparatus for making sense of the sea. As such, we too hope that the chapters gathered here help provide 'tools for refining the still crude categories through which scholars are attempting to apprehend seascapes, maritime histories, littoral cultures, and transoceanic exchanges' (Wigen, 2014: 12). In the case of IR, this conceptual rethinking concerns largely how we have conceptualised of time and space. With regards to the former, at its most basic, we need to rethink the way we conceive of the projection of power in light of the obstacles represented by the sea at different times. As to the latter, we need to review the extent to which we have taken ocean spaces as natural delimitators of space.

As 'spatial fillers', 'voids', or even 'the space between empires' (see the discussion above), the sea has tended to represent a vacuum between shores and acted as an invisible delimitator of space. For, as landmass is the primary object drawn by mapmakers, we have tended to ignore the extent to which the sea contributes to regional spatialities, to carving out masses of land and pushing them together. Yet the sea means different things to different people. For some, today, it is primarily a site of recreation – or even the primary site of recreation – while for others it may be an obstacle, a conduit, or a space for resource extraction. In a similar vein, the sea has not only worked to close regions off from one another, but also brought shores together. The Mediterranean is of course a case in point, but so is also the Baltic sea, while the English Channel has worked mainly to distance Great Britain from Europe. The spatial function of the sea therefore cannot be taken for granted but needs instead to be the object of our analyses.

In a similar vein, technological advances in seafaring proved key to the American Civil War, just as they did during the world wars. The importance of mastering the sea, then, has changed over time and cannot figure as a constant in our analyses. With the Cold War, for instance, it was argued that developments in intercontinental missile technology again made the sea rather irrelevant as one no longer had to master it in order to deliver a strike against any city on the globe. Yet again, this is a slight exaggeration as submarines still use the sea as hiding ground. All this goes to show that we cannot assume that the sea will play a passive role in our analyses. The sea needs to figure centrally in IR. But as we have emphasised above, bringing the sea in requires more than to study the sea. Studying the sea in isolation will not help us integrating the sea.

Rather, we have to think IR *with* the sea. In so doing, we need to understand how the different tropes which capture different ways in which the sea has been treated also cover different and often contradictory representations of the sea. We will illustrate this by way of a few examples. In so doing, we showcase what we mean by *thinking with the sea* and indicate why we have chosen the ensuing chapters for inclusion in the volume.

Taming or mastering the sea

The first trope is that of the sea as an untameable force, which works to contain human life to the shores. For a long period of time human offshore activity within and around certain seas was blooming, while the oceans represented vast and infinite space which could not be surmounted. Surmounting them requires both technical knowledge and know-how, but also a broad process of rethinking the significance of high seas. Perhaps the most basic intervention the authors of this book make in IR is that fact that mastery of the sea cannot be taken for granted, and that to master the sea requires a lot of resources. As Jan Glete has shown, state-sponsored seaborne military activity required much more resources than mounting terrestrial campaigns. In the modern era, this investment may be less onerous on states, yet that does not mean that the means invested in mastering the seas has become a marginal or insignificant. Maritime defence, even under the current hegemonic order we have regulating commerce and transportation over high sea lanes, requires vast resources and coordination.

Which brings us over to the relation between mastery of the sea and political domination. Through mastering sea lanes, one can control trade routes. A global mastery of sea lanes thus enables a broad control over global trade. As a case in point, the hegemonic role the British took from the second third of the nineteenth century would have been unthinkable without their mastery of the sea. The same could be said for the United States, which came to take over the role of global hegemon from World War II onwards. A global reach over the sea and firm control of central trade routes were also central in allowing the US to play the role they have played in the postwar world.

Thus, while the sea is something that is in the way, an obstacle to be surmounted or tamed, its mastery is also intrinsically connected to global dominance. Being a precondition for the establishment of empires beyond European landmass, the mastery of the sea is connected to power projection far beyond the immediate power of the vessels sent out. Investigating how global power is connected to the mastery of the sea, and how different forms of mastering the sea have engendered different understandings of power, rule and spatiality are key issues IR must address in order to understand the extent to which conceptions of order and relations between polities have developed in conjunction with changing technologies for mastering the sea. As long as the sea is an obstacle, the focus will be inland and towards regional politics, while the mastery of the sea – by turning the sea into a connecting space – turns the focus way from regionalisation and towards inter-regional interconnectedness. Mirroring this, we can hypothesise that our understanding of the sea works in a similar way on our theories of global ordering. Seeing the sea as a natural delimiter will guide us to see landmasses as ontologically given and make us prone to take those regions for granted – at the expense of connections across the sea which may in fact be stronger.

Traversing the waters

Most attempts to theorise IR on any macro scale today rest on an inability to conceive of a world without free circulation of goods. And this free circulation rests squarely on the sea. For the last two centuries, people and goods have been able to move at sea with fairly high certainty of not being molested. Never fully tested, this state of affairs has rested on British and American hegemony at sea, and potential rivals at least in principle adhering to common rules.

Once the obstacle of the sea has been mastered it represents a set of new challenges. It is also worth noting that not all seas have represented such a major challenge. In fact, certain sea lanes have been conduits of transportation and trade throughout history, shaping economic and political ties. As noted above, the Mediterranean comes to mind, as does the Caspian sea. The second trope of the sea as a conduit to be travelled, then brings out different effects of the sea, a sea that first and foremost connects. Such a view puts less emphasis on borders and what Mearsheimer, as discussed above, has called the 'stopping power of water' (2001) than on ties and connections. Focusing on the sea as a conduit or circulation of material goods, ideas and people, then, shifts the focus somewhat away from warfare onto trade, and away from states and onto greater political conglomerates such as empires and smaller network nodes such as ports.

Yet this way of approaching the sea tends to transform space to time. The sea is reduced to the time needed to traverse it. As the sea doesn't seem to represent a challenge, it is not focused upon either, and seems to gain importance more as a function of what travels on it than in and of itself. Even so, this trope has become increasingly important as the '[c]onnecting function of oceans has come to supercede their earlier distancing functions' (Wigen, 2014: 15), and therefore all the more pressing to focus on for IR.

Controlling maritime resources

The final trope the sea has been understood through is that as a repository for resources, resources whose extraction rests upon controlling parts of the sea. Fisheries and offshore oil drilling are obvious examples of how the sea appears, but the sea as a resource is not limited to that. Maritime states with long shores have more of a say in global sea matters. Long shores are also more 'territorial' than the high seas, as they fall within the exclusive control of maritime states. As such, the sea can convey authority to states. The artificial islands China is building in the South China Sea are a case in point, both in terms of how the sea and maritime resources can be controlled through processes such as territorialisation, akin to those processes taking place on land, but also in terms of how a state actively seeks the authority seashores convey upon states by actively building these shores.

The sea as a resource, then, draws our gaze to maritime contact zones, zones where sea and land meet. Just as control of the sea requires some relation to landed polities, extracting resources from the sea hinges upon one's claims of ownership or right to use the sea. While this places ports and other liminal spaces at the center of the analysis, it also draws in the framework within which these claims and rights are nested. International Law, especially the Law of the Sea (UNCLOS), is crucial in framing these claims and central to making sense of how the sea has been harvested in the past as well as today.

The question of maritime resources is of course also closely related to the extent to which these resources are available and how this has changed over time. The challenges posed by climate change are crucial in these terms, as for instance rising sea levels. Changes in the conditions of possibility of extracting resources and the changing nature of the sea/authority relation should be the focus of studies in IR, but are also ways in which a focus on the sea can contribute to IR more broadly. The interplay between land and sea has also become more important within the sphere of defence, where since the Battle of Midway (1942) and air power eclipsing traditional sea power, it has been obvious that shipping lanes could not be controlled solely by surface ships, but by an interplay between land, air and sea forces. A focus on the sea from the perspective of resources, then, also showcases the extent to which trade and violence are intertwined. In addressing maritime resources, IR scholars should focus on the extent to which the sea is changing (from fluidity to territorialisation?), and to what extent changes in global governance and the effects of climate change contribute to these changes.

Conclusion

The chapters of the book all address the sea through the tropes discussed above, making the case for why the sea ought to be more central to IR, and the implications of doing so beyond the subject matter at hand. Our goal has not been to cover the topic encyclopaedically, but to present agenda-setting and exemplary research. The chapters of this volume are nevertheless united by being theoretically informed and concerned with the historical developments of international relations and the sea; they explore how thinking historically and theoretically about international relations and the sea help open up the discipline.

Immediately succeeding this introduction, Alejandro Colás in Chapter 1 makes the case for understanding sea *and* land through their linkages, through 'terraqueous' relations and linkages. Following that, Maria Mälksoo draws our attention in Chapter 2 to the extent to which the sea is intertwined in the constructive of collective identities. In Chapter 3, Halvard Leira and Benjamin de Carvalho explore two dimensions of the sea which – even in studies of the sea – are largely marginalised, in spite of the extent to which they structure our understanding of it: gender and race.

Chapters 4 by Mark Shirk and Chapter 5 by Benjamin de Carvalho and Halvard Leira moves on from such broader theoretical engagements and delve into the historical importance of the sea in the constitution of our current global order. Both chapters focus on key developments in the politics of the sea which came to structure the make-up of the world beyond European shores. Dealing with seaborne or amphibious violence, both chapters remind us of the extent to which our current order rests on specific articulations of politics and law intertwined with maritime practices.

Dealing more explicitly with the present, in Chapter 6, Andonea Dickson analyses current practices around migration in the Mediterranean. In doing so, she addresses the sea as a geography, showing how through a multitude of contacts and entanglements, the sea and seafaring become central vantage points in defining political orders. In Chapter 7, Jessica Simonds continues to focus on the more practical side of traversing the sea, focusing on risks at sea, how they are perceived and how they are addressed in the context of the Indian Ocean. Focusing on the interplay between maps, navigational technologies and security, she addresses how threats to maritime transit become conceptualised as threats to landed political structures. Filippa Braarud continues this investigation of the relationship between the sea and modes of mastering it in Chapter 8, focusing on the deep seabed as common heritage of mankind. Her focus, drawing International Law explicitly into the argument, is on whether such protection withstand changes in technologies of extraction, and if these changes influence the extent to which the common heritage of mankind designation gives rise to *erga omnes* obligations. In Chapter 9, Kerry Goettlich emphasises the need for new ways of analysing the connections between humans and the environment. In attempting to take the environment seriously, Goettlich interrogates the extent to which the sea understood as the 'natural' world has played a crucial role in influencing core IR theories and categories.

In the conclusion of the volume, Xavier Guillaume and Julia Costa López draw up the implications of these interventions, emphasising the need to conceptualise the sea not as yet another space in which international politics takes place, but in terms of its many relationships with land, picking up the concept of terraqueous relations launched by Colás in the first substantial chapter of the book.

Together, these chapters seek to show the way in how IR could address the sea, and would be better off in doing so, as it would allow the discipline to recover the spatiality and politics of the sea. As Jordan Branch has shown, medieval mapmakers had no measure other than time to measure distance, as distance was measured in travel days. In medieval maps, space was measured in time. In a similar vein, we have argued here, the spatiality of the sea has been lost in IR, as space at sea has collapsed into time. In this focus on time, IR seems much like medieval cartography (Branch 2014), where the key items were places (often towns) and the time it took to pass from one of these to another, without much consideration to what happened between the 'places'. The spatiality of the sea has thus been overlooked, relegated to other disciplines, or assumed to be a corollary of land. This is striking, especially, perhaps, when it comes to liberal theories focusing on interaction and trade, since the vast majority of transport is seaborne. But in most liberal theories, oceans (if they appear at all) are simply reduced to time – the time it takes to cross them. Perhaps equally striking given the importance of mastering the sea as a precondition for global projections of power and the sea as a means to project landed power, the geographic imaginary of IR, formed by geopolitics, has centered around notions of heartland, rimland, etc. In fact, whatever IR has studied, it has been on land and not at sea.

Yet as we have argued here, the sea is a space, but also what creates spaces. As such, IR cannot remain blind to the sea and needs to think of the world with the sea rather than ignoring it. As the sea has determined where humans have settled, it has been largely responsible for our current human geography. Yet we still lack the most basic tools with which to grasp the sea. To quote (and paraphrase) Wigen, 'Most current categories of social analysis were initially developed to understand land-based societies. How those categories need to be transformed by perspectives from the sea – and how far they can be stretched, bent, and reworked to accommodate ocean-centered realities – is perhaps the most important unresolved agenda [ahead of us]' (2014: 17) This is the challenge this book seeks to address: start to think IR with the sea, and see where it takes us. The time has come to stop carrying the boats and to explore where the boats can carry us.

Notes

1 It also remains unclear whether the stopping power is related to medium or distance; but regardless, it is related to time – the time necessary to mobilise on an 'island', and the time necessary to cross to such an island.
2 It is impossible to overstate the importance of the work of Philip Steinberg for the growth of oceanic literatures over the last two decades. His seminal work on the social construction of the oceans (Steinberg, 2001) was published in an IR series, but had a much deeper impact in geography and adjacent disciplines. This volume could be read as a much belated IR follow-up to Steinberg's pathbreaking work.
3 While we find Schmitt's politics abhorrent, his panoramic vision of land and sea remains an important inspiration and a corrective to much 'landed' analysis. We nevertheless find his final conclusion unconvincing. Schmitt (2015 [1942]: 93) argues that: 'Today's transportation and communications technology has made the sea into a space in the contemporary sense of the word. […] if this is so, then the division of sea and land, upon which the link between sea domination and world domination allowed itself to be erected, falls away. The basis of British sea appropriation falls away, and, with it, what had up until now been the *nomos* of the earth.' On the contrary, we see a US continuation of British sea appropriation, and a continued division between sea and land, based on a continued perception that the sea is fundamentally different from land.
4 The notion of 'thinking with the sea' can be traced back at least to Rachel Carson's writing in the 1950s, and resonates with Lévi-Strauss' notion of 'being good to think with'. How the sea is 'good to think with' is obvious for example in the many references to 'flows', 'streams' and 'waves' in analyses of current affairs. We have obviously been unable to avoid them in this chapter as well. 'Thinking with the sea' has also been used as a grounding for 'wet ontology' (Steinberg and Peters, 2015). Our ambition here is, as noted, somewhat less ambitious.

References

Anderson, J., and K. Peters, eds (2014). *Water Worlds: Human Geographies of the Ocean*, Farnham: Ashgate.

Ashworth, L. M. (2011). 'Realism and the spirit of 1919: Halford Mackinder, geopolitics and the reality of the League of Nations', *European Journal of International Relations* 17(2): 279–301.

Barnes, T. J., and J. S. Duncan, eds (1992). *Writing Worlds: Discourse, Text and Metaphor in the Representation of Landscape*, London: Routledge.

Benton, L. (2011). *A Search for Sovereignty: Law and Geography in European Empires, 1400–1900*, Cambridge: Cambridge University Press.

Blakemore, R. J. (2013). 'British imperial expansion and the transformation of violence at sea, 1600–1850: Introduction', *International Journal of Maritime History* 25(2):143–5.

Branch, J. (2014). *The Cartographic State: Maps, Territory, and the Origins of Sovereignty*, Cambridge: Cambridge University Press.

Bueger, C., and T. Edmunds (2017). 'Beyond seablindness: a new agenda for maritime security studies', *International Affairs* 93(6): 1293–311.

Bueger, C., T. Edmunds and R. McCabe (2019). 'Into the sea: capacity-building innovations and the maritime security challenge', *Third World Quarterly* 41(2): 228–46.

Buzan, B., and G. Lawson (2015). *The Global Transformation: History, Modernity and the Making of International Relations*, Cambridge: Cambridge University Press.

Campling, L., and A. Colás (2018). 'Capitalism and the sea: sovereignty, territory and appropriation in the global ocean', *Environment and Planning D: Society and Space* 36(4): 776–94.

de Carvalho, B., and H. Leira (2021). 'State formation and historical International Relations', in B. de Carvalho, J. Costa López and H. Leira, eds, *Routledge Handbook of Historical International Relations*, London: Routledge, 231–43.

de Carvalho, B., J. Costa López and H. Leira, eds (2021). *Routledge Handbook of Historical International Relations*, London: Routledge.

Colás, A. (2016). 'Barbary coast in the expansion of international society: piracy, privateering, and corsairing as primary institutions', *Review of International Studies* 42(5): 840–57.

Colás, A. (2019). 'IR's sea sickness: a materialist diagnosis', paper presented at 6th European Workshops in International Studies (EWIS), Jagiellonian University, Kraków, Poland, 26–29 June 2019.

Colás, A., and B. Mabee (2010). 'The flow and ebb of private seaborne violence in global politics: lessons from the Atlantic World, 1689–1815', in A. Colás and B. Mabee, eds, *Mercenaries, Bandits, Pirates and Empires: Private Violence in Historical Perspective*, London: Hurst, 83–106.

DeLoughrey, E. (2017). 'Submarine futures of the Anthropocene', *Comparative Literature* 69(1): 32–44.

DeLoughrey, E. (2019). 'Toward a critical ocean studies for the Anthropocene', *English Language Notes* 57(1): 21–36.

Epstein, C. (2008). *The Power of Words in International Relations: Birth of an Anti-Whaling Discourse*, Cambridge, MA: MIT Press.

Gillis, J. R. (2013). 'The blue humanities', *Humanities* 34(3), www.neh.gov/humanities/2013/mayjune/feature/the-blue-humanities (accessed 14 June 2021).

Glete, J. (2000). *Warfare at Sea, 1500–1650: Maritime Conflicts and the Transformation of Europe*, London: Routledge.

Jackson, S. E. (1995). 'The water is not empty: cross-cultural issues in conceptualising sea space', *Australian Geographer* 26(1): 87–96.

Klein, B., and G. Mackenthun, eds (2004). *Sea Changes: Historicizing the Ocean*, New York and London: Routledge.

Lambert, A. (2018). *Seapower States: Maritime Culture, Continental Empires and the Conflict That Made the Modern World*, New Haven, CT: Yale University Press.

Langewiesche, W. (2004). *The Outlaw at Sea: A World of Freedom, Chaos and Crime*, London: Granta.

Law, J. (1986). 'On methods of long-distance control: vessels, navigation and the Portuguese route to India', in J. Law, ed., *Power, Action and Belief: A New Sociology of Knowledge?*, Abingdon: Routledge, 234–63.

Leira, H., and B. de Carvalho (2011). 'Privateers of the North Sea: at world's end – French privateers in Norwegian waters', in A. Colás and B. Mabee, eds, *Mercenaries, Pirates, Bandits and Empires. Private Violence in Historical Context*, New York: Columbia University Press, 55–82.

Lobo-Guerrero, L. (2011). 'Insurance, climate change, and the creation of geographies of uncertainty in the Indian Ocean region', *Journal of the Indian Ocean Region* 6(2), 239–51.

Lobo-Guerrero, L. (2012). 'Lloyd's and the moral economy of insuring against piracy: towards a politicisation of marine war risks insurance', *Journal of Cultural Economy* 5(1), 67–83.

Lobo-Guerrero, L., and Anna Stobbe (2016). 'Knots, port authorities and governance: knotting together the Port of Hamburg', *Global Society* 30(3), 430–44.

Mack, J. (2011). *The Sea: A Cultural History*, London: Reaktion.

Mahan, A. T. (2018 [1890]). *Influence of Sea Power upon History, 1660–1783*, London: Pamtianos Classics.

Mearsheimer, J. J. (2001). *The Tragedy of Great Power Politics*, New York: Norton.

Mentz, S. (2009). 'Toward a blue cultural studies: the sea, maritime culture, and early modern English literature', *Literature Compass* 6(5): 997–1013.

Modelski, G., and W. R. Thompson (1998). *Seapower in Global Politics, 1494–1993*, Basingstoke: Macmillan.

Mulich, J., and L. Benton (2015). 'The space between empires: coastal and insular microregions in the early nineteenth-century world', in Paul Stock, ed., *The Uses of Space in Early Modern History*, Basingstoke: Palgrave Macmillan, 151–71.

Ogborn, M. (2002). 'Writing travels: power, knowledge and ritual on the English East India Company's early voyages', *Transactions of the Institute of British Geographers* 27(2), 155–71.

Paine, L. (2013). *The Sea and Civilization: A Maritime History of the World*, New York: Vintage.

Peters, K. (2010). 'Future promises for contemporary social and cultural geographies of the seas', *Geography Compass* 2(6), 1260–72.

Peters, K., and P. Steinberg (2019). 'The ocean in excess: towards a more-than-wet ontology', *Dialogues in Human Geography* 9(3): 293–307. https://doi.org/10.1177/2043820619872886.

Phillips, A., and J. C. Sharman (2015). *International Order in Diversity: War, Trade and Rule in the Indian Ocean*, Cambridge: Cambridge University Press.

Rosenboim, O. (2017). *The Emergence of Globalism: Visions of World Order in Britain and the United States, 1939–1950*, Princeton, NJ: Princeton University Press.

Schmitt, C. (2015 [1942]). *Land and Sea. A World-Historical Meditation*, Candor, NY: Telos Press.

Shilliam, R. (2015). *The Black Pacific: Anti-Colonial Struggles and Oceanic Connections*, London: Bloomsbury.

Sivasundaram, S., A. Bashford and D. Armitage (2018). 'Introduction: writing world oceanic histories', in D. Armitage, A. Nashford and S. Sivasundaram, eds, *Oceanic Histories*, Cambridge: Cambridge University Press, 1–27.

Steinberg, P. E. (1999). 'Lines of division, lines of connection: stewardship in the world ocean', *Geographical Review* 89(2): 254–64.

Steinberg, P. E. (2001). *The Social Construction of the Ocean*, Cambridge: Cambridge University Press.

Steinberg, P. E. (2014). 'Foreword: on thalassography', in Jon Anderson and Kimberley Peters, eds, *Water Worlds: Human Geographies of the Ocean*, Farnham: Ashgate, xiii–xvii.

Steinberg, P., and K. Peters (2015). 'Wet ontologies, fluid spaces: giving depth to volume through oceanic thinking', *Environment and Planning D* 33(2): 247–64.

Subrahmanyam, S., ed. (1996). *Merchant Networks in the Early Modern World*, Aldershot: Ashgate.

Till, G. (2018). *Seapower: A Guide for the Twenty-first Century*, 4th edn, London: Routledge.

Tilly, C. (1992). *Coercion, Capital, and European States, AD 990–1992*, London: Wiley-Blackwell.

Trim, D. J. B., and M. C. Fissel, eds (2006). *Amphibious Warfare 1000–1700: Commerce, State Formation and Expansion*, Leiden: Brill.

Westerdahl, C. (2005). 'Seal on land, elk at sea: notes on and applications of the ritual landscape at the Seaboard', *Nautical Archaeology* 34(1): 2–23.

Wigen, K. (2014). 'Introduction', in J. H. Bentley, R. Bridenthal and K. Wigen, eds, *Seascapes: Maritime Histories, Littoral Cultures, and Transoceanic Exchanges*, Honolulu: University of Hawai'i Press, 1–18.

Winkiel, L. (2019). 'Introduction', *English Language Notes* 57(1): 1–10.

1

International Relations' sea sickness: a materialist diagnosis

Alejandro Colás

Since the formal inception of International Relations (IR) as an academic discipline after World War I, only a handful of significant scholars have addressed the place of the sea in international relations.[1] Schematically, these have fallen into either the geopolitical Realism of Haushofer (1938), Schmitt (1997), Spykman (1944) and – in a different register – Mearsheimer (2014), or the liberal institutionalism of Peter Haas (1990) and Oran R. Young (1992). Despite their radically contrasting political outlooks, the sea has remained a relatively static space in these accounts of world politics – acting as either a passive backdrop in the exercise of terrestrial authority, or as a mere geographical void that represents the 'stopping power of water' (Mearsheimer, 2014). More recently, the critical-geographical work of Phil Steinberg (2001), and a new wave of historical-sociological accounts of the modern states system in part inspired by Lauren Benton's 'legal pluralism' have, together with 'tidalectical' interpretations from the Pacific and Caribbean, begun to recognise the oceans as distinctive spaces of world politics (e.g. Shilliam, 2015). Phil Steinberg and Kim Peters (2015: 257) have perhaps gone furthest in invoking a 'wet ontology' which looks at the ocean 'not as a space of discrete points between which objects move but rather as a dynamic environment of flows and continual recomposition [or "churning" as they elsewhere put it] where, because there is no static background, "place" can be understood only in the context of mobility.' They advocate an incorporation of the oceans as vibrant matter that interacts chaotically with other planetary vital forces to generate 'new understandings of mapping and representing; living and knowing; governing and resisting' (p. 258), i.e. disruptive, fluid and de-centred conceptions and practices of (international) politics.

Though not as wedded to the notion of 'assemblage' that accompanies Steinberg and Peters' wet ontology, Andrew Phillips and Jason Sharman's (2015) work on heterogeneity in the Indian Ocean international system, or Jeppe Mulich's (2015) incursions into the coastal 'space between empires' also enjoin us to incorporate the diversity or plurality of polities and territorialities fostered in large measure by the particularities of the sea into our explanations of international relations. Such ocean-facing approaches to world politics differ from more 'idealist' conceptions, one recent and sophisticated example of which is Andrew Lambert's (2018) theory of seapower states. For Lambert, the latter represent particular forms of oligarchic rule which – from Athens and Venice through to the Dutch Republic and Great Britain – fashioned inclusive, dynamic, outward-looking and progressive polities and cultures with the sea as their chief commercial and diplomatic resource. In direct contrast to the geographical-historical materialism outlined below, Lambert argues that the sea is mobilised as site of a maritime culture in the creation of seapowers: 'it is not a consequence of geography, or circumstance. The creation of seapower identities has been deliberate, and is normally a conscious response to weakness and vulnerability' (2018: 8).

There is, then, a growing and engaging body of work within contemporary IR that takes on board the importance of distinctive political ecologies upon diverse dispensations of power between political communities – that is, from a maritime perspective, a view which considers the saltwater part of our planet as a dynamic, changing and differentiated force in world politics. This furthermore generally implies adopting some kind of materialist understanding of nature as having distinguishing properties that are unevenly transformed through human intervention. With regard to the sea, its fluidity, salinity, depth, density, biomass and energy represent some of these unique natural attributes (although of course seawater can also solidify into ice and evaporate into air), and the oceans' hostility towards permanent occupation its social distinctiveness. In *The Nomos of the Earth* and in his earlier essay *Land and Sea*, Carl Schmitt posited an ontological division between static landmasses and fluid oceans where the former deliver an order built on law, delimitation and appropriation, and the latter is a sphere of open, borderless, common ownership – the double vowels in its German rendition (*Meer*) evoking an characterless void; a transitory space merely defined in relation to land. In contrast to land, where the triple process of appropriation-distribution-production (or more literally 'seizing-dividing-tending') aligns this spatial orientation to a concrete order, the ocean space is a featureless realm of freedom and universality: 'on the sea, fields cannot be planted and firm lines cannot be engraved' (Schmitt, 1997: 1).

One need not fully accept Schmitt's telluric metaphysics (even less so the odious politics that accompanied them) to acknowledge that the oceans' resistance to effective occupation in the way that political sovereignty is imposed on land gives the blue water world a distinctive power in international relations. The rest of this chapter draws out the implications of this line of thinking by offering an overview of what a terraqueous conception of the world can add to the study of IR. For, glaringly self-evident as it is, the multiple socio-economic and political implications of the earth's separation into land and sea have largely been overlooked in IR, suggesting the 'malaise' of the discipline when it comes to the maritime factor in international relations lies in the lack of attention to this elementary geophysical interaction. Exploring world politics through a terraqueous lens, I suggest, deepens and enriches our understanding of international relations in multiple ways.

Ordering the unruly sea

The first of these concerns the challenges of imposing order on a high sea that is nobody's property. During the early modern period, England and the Netherlands in particular made the oceans a prized venue not just of state formation through plunder, profit and naval prowess, but also of world-ordering via new doctrines of international law. 'Freedom of the Seas' (*Mare Liberum*) was one such principle, famously articulated by Hugo Grotius under instructions of the Dutch East India Company, as the latter sought to secure its seaborne trading monopolies (*dominium*) while insisting the high seas were exempt from possession by any single jurisdiction (*imperium*). The tight public–private partnership between political rulers and trading companies which characterised both Dutch and English fiscal-military state (in seventeenth-century Netherlands, its

East India Company *was* the state) reflected the centrality of maritime commerce to the very existence of these rising powers (Brandon, 2015). Thus, at the very least, we should continue with recent historiographical critiques decentring and destabilising the Westphalian foundational myth by adding the Malacca Straits – where the privateering incident that prompted Grotius' *Mare Liberum* unfolded – as a synchronous birthplace of the modern international system (see Anievas and Niçancloğlu, 2015). We might then also think of modern territorial sovereignty as an amphibious affair that witnessed new socio-economic and legal-political forms and processes generated by the unique interaction between land and sea.

A good illustration of the materiality of such terraqueousness in shaping modern international relations lies in the practice of piracy. Predation on the high seas is as old as war and trade but, as de Carvalho and Leira's contribution to this volume shows, it acquired unique form and function in the age of mercantile empires, both as (legitimate) privateering and (outlaw) piracy. In particular, maritime predation both exploited and subverted freedom of navigation as state-sponsored and non-state actors alike preyed freely on enemy vessels, often recycling personnel, ships and letters of marque and reprisal across public and private domains (Rediker, 1987). Despite parallels between land-based banditry and warlordism on the one hand, and seaborne piracy and privateering on the other, the ocean's fluid nature dilutes attempts at establishing a monopoly over the means of violence across its surface. Public authorities, including most recently multilateral organisations like the UN or international alliances like NATO, have tried to enforce the law through 'transit corridors' or 'high risk zones' in the western Indian Ocean. But it is telling that the preferred spatial forms here are undelimited zones rather than bordered territories, and that much of the counter-piracy monitoring and surveillance is conducted from onshore sites. Similar amphibious experiments characterised the early eighteenth-century war on Atlantic piracy, which relied on a necklace of colonial outposts across the region – military garrisons, trading entrepots, diplomatic enclaves – both supportive of, and protected by, the Royal Navy's anti-buccaneering campaigns, thereby generating what Lauren Benton memorably called the 'imperfect geographies and … variegated spaces' of a terraqueous legal pluriverse (Benton, 2010: 2).

By the same token, an unintended consequence of the eighteenth-century wars on piracy was the development of universal jurisdiction against the figure of the pirate as an 'enemy of humankind' (*hostis humani generis*). This legal norm allowed ship captains of any flag to apprehend and, in extreme cases, court-martial and execute those engaged in acts of piracy on the high seas – a principle that was also subsequently extended to slaving ships once the trade in humans was outlawed. It has more recently informed notions of humanitarian law and crimes against humanity which trump principles of sovereign immunity or territorial integrity when addressing mass violations of human rights. The refractory nature of the sea has therefore engendered innovative forms of international rule, most notably though not exclusively through law and governance which, in instances like universal jurisdiction, have found their way back onto land. The oceans thus act as an obstacle that requires taming or mastering, as mooted in the book's introduction, in the process producing specific expressions of international relations. The challenge for those wishing to render the bluewater world visible in IR is to identify these specificities: what is unique and distinctive about the relationship between land and sea at any given time and in any particular place?

One response is to consider, as Phil Steinberg (2001) did in his seminal volume on the subject, different social constructions of 'ocean governance' – ranging from the 'asocial' conceptions of the Indian Ocean as 'a space between societies', to the 'stewardship' model characteristic of both Mediterranean antiquity and modern European merchant empires, where the sea acts as a 'force field' that can be controlled (e.g. through protected sea lanes and convoying) but not possessed. From a materialist perspective, there is much to recommend in this understanding since it underlines the centrality of social reproduction when identifying the particularities of different socio-natural relationships between land and sea. Plainly, the 'deep time' of our bluewater planet has allowed for, and indeed conditioned the development of, multiple geopolitical relationships between land and sea across history. Barbadian poet and intellectual Kamau Brathwaite identified a diasporic 'submarine unity' among African peoples forcibly transported from their homeland, and such underwater connections also apply to other sea-going civilisations, both during and before the modern period (Brathwaite, 1974: 64). Several millennia before the Roman Empire constructed its famed road network, the coastal peoples of Atlantic Europe had established a dense lattice of 'seaways' – the Gaelic *astar mara*, the Norse *veger* or *hwael-weg* (the whale's way) in Old English – which linked today's northern Spain with England, France, Ireland, as well as the Atlantic and Baltic beyond, while the Norse and Icelandic sagas reflect a rich culture of maritime trade, skill, plunder, conquest and enslavement where the oceans become an active historical force (Macfarlane, 2011: 91). Seafaring populations of the Pacific have, moreover, for centuries developed intricate cosmologies that conceive of the waters surrounding them as a 'sea of islands'. Far from being an alien and alienating space beyond everyday human experience, the sea was for most Pacific peoples a sacred ancestral resting place as it was a source of protein and propulsion (Lewis, 1994).

In western Europe on the other hand, as Alain Corbin's (1994) classic study suggests, the ocean was mainly feared and avoided as a realm of chaos and destruction until the ascent of Dutch maritime hegemony in the early seventeenth century, when the sea and its shores started to be the subject of admiration. To be sure, as in the rest of the world, pre-modern distant-fishery communities (including Basque, Breton and Norse 'nations') could be found across Europe's Atlantic littoral, their folklore and worldviews deeply interconnected with the sea. But the bulk of coastal populations in Europe combined near-shore fishing with inland agriculture, thereby replicating an almost universal suspicion of, when not outright disdain for, merchants and fear of 'going out' to sea. Even classical and modern empires – from ancient Rome to the Ottomans – generally mobilised their navies for purposes of conquest or commerce (the Ming Dynasty abruptly cancelling all Chinese maritime expeditions in 1433), rather than gearing their societies towards wealth accumulation through trade. Of course, these and many other continental empires engaged in overseas trade and plunder, but this was ancillary to their dominant mode of social reproduction premised on the extraction of tax and tribute from those living on land. Put bluntly, a tributary empire like Habsburg Spain used the sea as a medium to extend its terrestrial frontier; merchant empires like seventeenth-century Netherlands on the other hand used their territorial sovereignty to control lucrative sea lanes. The former

extended to the Americas the practices of the Iberian raiding frontier, essentially extracting precious metals from subject populations;
the latter exploited commercial networks by facilitating and intensifying the maritime circulation of commodities from one marketplace to another.

Another, by no means incompatible, response to the question of
what is particular to the spatio-temporal relationship between land
and sea is to recalibrate the conventional temporalities of international relations according to the geophysical properties of the sea.
That is, to re-envision notions of world-time through experiences
of 'the international' *at* sea, and created *by* the sea. As just noted,
the long sixteenth century conventionally marks the beginning of
a modern epoch which generated the agents and structures – the
sovereign state, the global market, social classes and ethno-national
identities among others – that continue to drive international
relations today. Yet accompanying these generally more visible
forces of epochal change are historical undercurrents literally
submerged below the sea's surface which offer counternarratives
to modern IR, or at least afford perspectives that enhance and
complement prevailing conceptions of the modern world system.
The much-quoted poem 'The Sea is History' by St Lucian Nobel
Laureate Derek Walcott has become a shorthand for these complex and overlapping conceptions of historical time that shadow
modernity's relationship to the sea. From the perspective of many
Americans and Caribbeans of African descent, the answer to the
loaded question 'but where is your Renaissance?' is: 'Sir, it is locked
in them sea-sands/out there past the reef's moiling shelf, where the
men-o'-war floated down' (Walcott, 1978: 137–9). Contrary to
the common perception of the sea as a smooth horizon of opportunity which simply connects one market to another, merely acting
as a surface of circulation, the ocean seabeds – not just of the
Atlantic – can also be seen as the underwater resting place, both
real and imagined, of lives sacrificed and destroyed at sea.

The Atlantic slave trade is one notable instance of this. The
reinvention of chattel slavery in the Western hemisphere integrated
the merchandise of captive Africans into the emerging seaborne
world market, thereby contributing to the process of capitalist
development in Europe. The sea played a key role in this experience, acting both as the main conduit for human trafficking on an

industrial scale, and as the site for the production of geographical distance necessary in the generation of profit through conveyance. The Atlantic Ocean thus became a real geophysical and logistical barrier in forcible population transfers, while also creating multiple profit-making opportunities for insurers, traders, bankers, slavers, shipbuilders and factory-owners by turning distance into a market for risk, credit, manufacture, commodity exchange and transport – realising differential accumulation when 'buying cheap' in one coast and 'selling dear' in the other. Viewed from the depths of the Atlantic and the holds of the slave ships that criss-crossed it, there is therefore something historically peculiar about the modern ways of social reproduction at and through the sea.

While the rise of Atlantic slavery signals the sixteenth-century conjuncture that inaugurated the modern international system, the place of the ocean world in today's unfolding climate emergency points to a deep time that has moulded our world in the very *longue durée*. The sea and its islands are a repository for all kinds of terrestrial waste, excess and surplus, the most threatening of which is the amount of carbon and heat absorbed by the oceans. Nowhere is the presence of historical capitalism's production of CO_2 and other greenhouse gases (GHG) more apparent than in the heating, expansion and acidification of the oceans. If the compound accumulation of past GHG emissions today means, in Andreas Malm's (2016) arresting phrase, that 'the air is heavy with time', even more so is the sea. Indeed, the widely contrasting temporalities at play here – marine life has been slowly changing over millions of years, only to be rapidly transformed in the space of a few decades by a social system barely a few centuries old – already tell a powerful story about the disjointed timescales of the crisis we are facing. To make matters worse, the centrality of the oceans to the reproduction of our biosphere – its critical role in regulating atmospheric temperatures and humidity, and thereby extreme weather events; its significance in the global food chain; its part in sustaining both marine and terrestrial ecosystems – accentuates the risks of seaborne climate change. The full implications of warmer, anoxic and acidified oceans today may not become entirely apparent until a more distant future since the sea's heat and carbon release is a more protracted process than, say, that produced through deforestation.

Incorporating the sea and its socio-natural interaction with land into an analysis of IR thus allows us to articulate specific events with conjunctural and deep time, arguably enriching our causal assessment of change and continuity in world politics. A good example of this is the Indian Ocean monsoon system, which combines this tripartite division of historical time in shaping (and indirectly, being shaped by) not just the circuits of seaborne commercial, spiritual and cultural exchange across that intercontinental basin, but also the political ecologies of much of South and Southeast Asia (see Risso, 1995; Subramanian, 1999; Sheriff, 2010; Cederlöf, 2013; Prange, 2018). Time is inscribed in the very etymology of the word 'monsoon' (derived from *mawsim*, Arabic for 'season') and its existence for millions of years as a recognisable weather pattern integrated into the planetary El Niño/Southern Oscillation attests to the permanency of the monsoon in the long duration. The fact that it is the 'thermal contrast between the land and the ocean, and the availability of moisture' that drives monsoons and determines their socio-natural effects moreover underscores the terraqueous character of the phenomenon (Amrith, 2019). Across 'Monsoon Asia', peasants and sailors, rural farmers and urban dwellers have through the centuries learnt to read the skies in anticipation of either much-needed wind and rainfall, or catastrophic typhoons and flooding (sometimes all of the above).

Global warming is manifestly disrupting and altering monsoons in ways that make their accompanying weather patterns more erratic and extreme. Yet the temporal disjuncture between the immediate causes and future consequences of climate change makes it difficult to pinpoint these radical changes to any specific event in any particular place. There is no clear beginning or predictable political outcome to climate change, nor a single, identifiable enemy that might be defeated (although plainly decarbonising our societies might begin to mitigate the climate crisis). Instead, there are a multiplicity of causes and temporalities that have to be disentangled and then reassembled into a tentative whole. Here, focusing on the terraqueous nature of the monsoon across an event in time, a historical conjuncture, and the long duration of deep time gives us greater purchase on the current predicament and the past sequencing of critical aspects of world politics, like the growth of coastal megalopolises, the intensifying inequalities within and between urban and rural livelihoods, or the disappearance of small island states resulting from sea-level rises. This is emphatically not about reverting to some spurious environmental determinism where geography is destiny, but it does require giving due causal power to the natural forces particular to the relationship between land and sea – that is, taking seriously the contribution of the ocean's unique material properties to the evolution and transformation of the international system.

Ocean crossings

That much of the history of international relations concerns taming the seemingly unruly high seas should not distract us from the blindingly obvious fact that the oceans have mainly served as facilitators of movement – as the fastest and relatively cheapest highway for all sorts of intercontinental transactions between peoples and nations (Stopford, 2009). Traversing the sea has multiple implications for human societies, all of which involve some reckoning of geographical distance through distinctively maritime instruments of measuring and regulating space and time. What Jason W. Moore has labelled the 'technics of global appropriation' (the caravel, the magnetic compass, the seaman's astrolabe, and their corresponding maps and charts) clearly involved attempts at ordering the sea during the modern period, as we just saw (Moore, 2015). But these instruments needed to be designed, manufactured, mastered and operated by those with seafaring knowledge and expertise. This also applied to the seamen, doctors, cooks and – after the age of sail – stokers, trimmers and engineers who made long-distance shipping possible. The very notion of 'globalisation' and its associated practices is therefore intimately connected to the sea, and deepening our conception of international relations so as to encompass diverse transnational phenomena necessarily involves paying attention to lives spent at sea. A brief consideration of maritime logistics can shed light on ways in which the geophysical properties of the sea, and its relationship to land, have underpinned some of the most salient practices and institutions of modern international relations, including 'global governance'.

The modern shipping container, also known as the Twenty-Foot Equivalent Unit (TEU), is the socio-technical artefact most often linked to seaborne globalisation. As Rose George and Marc Levinson's celebrated books would have it, by transporting '90% of everything' the container is the box that 'made the world smaller and the world economy bigger' (George, 2013; Levinson, 2006). The universal equivalence designed into the shipping container, as well as its inter-modal quality as a 'sea-going truck', has made the TEU an especially loaded metaphor for a capitalist globalisation where the ocean freeways enable the annihilation of space by time (Bonacich and Wilson, 2008; see also Klose, 2015). Yet such Whiggish associations of seaborne commerce with progress and prosperity tend to underplay how much logistics is above all a form of capitalist *planning* – with all its socio-political frictions and uncertainties – and how far this 'art and science of management' originates in the imperial age of steamship lines (with notable antecedents in the Atlantic slave trade).[2] The sea's critical role as a flat, horizontal transport surface thus has to be complemented by an appreciation of the very vertical hierarchies that accompany the unfolding of maritime logistics – including those of state agencies, labour processes, racist laws, and indeed geophysical phenomena ranging from storms and currents to water depth and ice extent. Much of modern international relations, even more so the global political economy, has been shaped by these various social and natural forces issuing from the sea.

The essence of capitalist logistics lies in managing turnover time so that each transfer across any given supply chain is completed as smoothly and cost-effectively as possible. This usually means accelerating movement so as to minimise turnover time, but it can also involve cutting labour and capital costs ('slow steaming' of ships is one way of reducing fuel outlays in maritime logistics) or investing in new time-saving and perishability-reducing technologies. Reliability and regularity are the main currency of logistics; planning and coordination the way of securing market share and profitability. In principle, shipping has the comparative advantage of uninterrupted transit through the 'free sea'. Yet, even if weather, piracy, war or technical breakdown fail to disrupt navigation, different degrees of land-side 'friction' (customs inspections and paperwork, delays, strike action, repairs and, in extreme cases, war and conflict) can and do affect the management of turnover time (Cowen, 2014). Interstate agreements, multilateral cooperation, industry and regional development bodies, port authorities and shipping firms, international and maritime law, and sector-wide collective bargaining (i.e. state, capital, labour and international organisations) will all determine the nature and efficiency of such logistical planning. Moreover, the terraqueous features of our planet make an appearance once more as shipping routes are still largely dictated by enduring geophysical features – aside from the Panama and Suez canals, most of the world's maritime chokepoints remain the same as in previous millennia.

Sea merchants have been aware of such spatio-temporal determinants of their trade since time immemorial, but the advent of capitalist shipping – and the introduction of industrial steamships in particular – witnessed the commercial decoupling of cargo ownership from the transport of commodities (and indeed shipbuilding), thus intensifying the separation between traders, shippers and shipbuilders (Paine, 2015). This in turn encouraged during the second half of the nineteenth century the frenzied development of a global logistical infrastructure in docks, wharfs, coaling stations, passenger ports, freight forwarding roads and railways, as well as the accompanying undersea telegraphic networks which sustained such intercontinental webs of transport and communication (Müller and Tworek, 2015). Imperial power was at the root of such technological globalisation, and liberal internationalism its outgrowth. Seen in this light, it is perhaps unsurprising that some of the first multilateral institutions of 'global governance' had early origins in European and world conferences dealing with riverine navigation (1861, 1863 and 1866), marine signalling (1864), and the 'neutralisation of submarine cables' (1882) (Murphy, 1994: 57). The liberal internationalist romance with the ocean as a space that underwrites progress, prosperity and liberty (perhaps most emblematically represented in Captain Nemo's proclamation that 'the sea does not belong to despots') thus reflected a reality sustained by the *Pax Britannica* (1815–70), where millions could pursue across a seaborne world market what Marx sarcastically referred to as the dream of 'Freedom, Equality, Property and Bentham' (Verne, 2014: 66). The Grotian promise of the free seas was here married to the Kantian notion of international communication and cooperation delivering perpetual peace.

Of course, these experiences of maritime internationalism tell only part of the story of the nineteenth century's 'great transformation'. For in the lower decks of ships; in the imperial wharves, dockyards and naval bases dotted across the globe, a saltwater cosmopolitanism was also being forged among the subaltern classes of all races and nations who served as the 'muscles of empire', both onboard and ashore (the phrase is Frank Broeze's, 1981). Maritime logistics above all requires manpower (though women have, like in the rest of society, historically played a crucial reproductive role in seafaring lives) (Creighton and Norling, 1996). Shifting bulk across oceans is almost by definition reliant on multinational workforces drawn from diverse labour markets. This was (and in many instances continues to be) in part a consequence of chronic labour shortages occasioned by, inter-alia, protectionist laws, wartime demands on navies, and the dangerous, gruelling, confined and therefore unappealing nature of work at sea, generally compounded by the high mortality and desertion rates, particularly among European crew on outgoing voyages. In the course of the nineteenth century, internationalising the workforce also became part of a deliberate cost-saving and labour-disciplining strategy whereby ratings were hired, paid and organised along a racialised hierarchy reliant on labour conveyancing (recruiting low-waged labour in foreign ports) and the crew management of intermediary 'headmen' (Fink, 2014). Underlying these practices is the ocean's unique quality as a workspace that produces geographical distance, yet is premised on the physical proximity of crewmembers in the highly stratified and bounded 'mobile factories' that are ships.

If the sea is a site for the formal expressions of international relations in the shape of international law, global governance or multilateral cooperation, it is also a venue for international relations understood as the more prosaic movement of people, commodities and ideas across states. The figure of the 'lascar' – a seaman originating from any part of the Indian Ocean region hired to work on metropolitan ships – is perhaps the quintessential personification, together with the eighteenth-century 'motley crew', of this contradictory combination of movement and confinement, freedom and domination, anarchy and hierarchy that encapsulates both types of international relations.[3] Lascars were simultaneously essential to the reproduction of empire into the early twentieth century, and an oppressed, marginalised class of colonial subjects; their ethno-national identity was indeterminate and fluid, yet their status as 'lascars' sharply codified and violently enforced. Such racialised and super-exploitative conditions continue today among the bulk of the world's ocean-faring workforce (Couper et al., 2015). However, as in the past, the singularly strategic position of the bluewater proletariat in the global economy, together with its inherently multinational composition, offers unique opportunities for inter-nationalist politics (Sampson, 2013). As in the case of universal organisations, it is unsurprising that the late nineteenth-century 'new' mass unionism was pioneered by workers in maritime-oriented sectors around, for instance, the International Transport Workers' Federation, ITF (though, anecdotally, the modern usage of 'strike' in industrial relations originates in London-based sailors and port-workers 'striking the sail' for pay rises in 1767) (Broeze, 1991). Nor is it a mystery why stevedores or longshoremen have historically been at the forefront of internationalist solidarity – with fellow workers or against repressive regimes like apartheid South Africa (see Weiss, 1996, 2017; Featherstone, 2015; Alimahomed-Wilson and Ness, 2018). These same ocean-facing workers – dockers, shipbuilders, seamen – have, however, also represented the most egregious instances of labour aristocracies using trade unions as racist closed shops organised along strict colour lines. The inher-ently globalising dynamics of maritime logistics, and all the labour that subtends it, therefore represent the critical place of ocean crossings in the construction of modern international relations.

Valorising the oceans

We have thus far seen how the high seas present both challenges and opportunities to diverse transnational maritime agents, in the process generating new practices, institutions and structures of inter-national relations. At stake here is a particular dialectic between land and sea, where socio-natural forces clash (and occasionally cooperate) over the shape of order, movement, commodity circula-tion and sovereign rule across our terraqueous planet. These conflicts and contradictions are resolved, however temporarily, through new global configurations of power mediated through law, multilat-eral agencies, maritime corporations, and the labour organisations representing saltwater proletarians who keep the world economy moving. The sea's unique geophysical properties, moreover, open fresh possibilities for rethinking the conventional temporalities prevalent in IR – be these related to the periodisation of our contemporary international system, or the deeper notions of ecological time and how they connect to biospheric crisis. The material power of the oceans is reflected in a further arena of international affairs, namely the appropriation of the ocean's energy and biomass.

Two of the notable legal-political innovations in accommodating the sea to a logic of territorial sovereignty are Exclusive Economic Zones (EEZs) and Open Registers (Flags of Convenience, FOC). Although emerging at different junctures in the last century, and often propelled by divergent constellations of interests, both these juridical forms reveal how the recalcitrant qualities of the sea in the face of attempts at imposing terrestrial methods of rule have in fact delivered novel and peculiarly hybrid modes of terraqueous territoriality. The EEZ is emblematic of such spatial effects in that it incorporates sovereignty (exclusive), appropriation (economic) and territory (zone) in its very title. The codification of the EEZ under UNCLOS III in 1982 was the single greatest territorial enclosure in human history. The outcome was a distinctive legal framework allowing coastal states to claim special sovereign rights (but not territorial sovereignty) over a delimited zone 200 nautical miles from a defined littoral baseline.

Under this order, so long as the principles of freedom of navigation and innocent passage for the world's largest fleets are upheld, coastal states can do with their marine resources as they please. The sea's currents and the biomass that move with them, however, do not respect the tidy logic of the states system. Highly migratory species like tuna in particular continually, and naturally, subvert any straight lines mapped upon the ocean space: as such, straddling fish species can only be nobodies' property until they are caught. The socio-spatial form of the EEZ thus challenges a common view of the global ocean as a lawless frontier. Even on the high seas, fishing activities are governed by complex layers of international law, including the partial regulatory reach of regional fisheries management organisations over fish stocks, the International Maritime Organization's authority over shipping pollution, and the International Labour Organization's remit over the pay, working conditions, and occupational health and safety of crew on boats (DeSombre, 2011). For its part, an International Seabed Authority administers the ocean floor beyond extended continental shelves (the 'Area') as the common heritage of humanity (including for deep-sea mining and bioprospecting), while the self-explanatory Commission on the Limits of the Continental Shelf does something similar when setting the baselines of territorial waters (Hannigan, 2016).

There are, then, plenty of instances where diverse global governance regimes and institutions seek to manage the global ocean in its surface, deep-water and sub-sea totality. The difficulty for many of these multilateral agencies lies in conjugating the liberal principle of the 'freedom of the seas' with the drive to secure sovereign property rights over, and the capture of ground-rent through, these resources. The EEZ represents one such attempt at marrying unfettered mobility and legal appropriation, albeit with the sea in this context serving as a laboratory in the experimentation with forms of overlapping governance that have subsequently been applied on land. In this regard, it is helpful to understand the EEZ not just as an area or zone but also, as Gavin Bridge has suggested, to consider it in volumetric terms 'as a spatial form of property through which the circulation of resources and commodities is controlled'. In contrast to a stable, purely grounded conception of resources as 'fixed territory', Bridge encourages us to think of 'quanta-based' rights to fish, water or other biomass as the principal way that capital can 'secure flow' (Bridge, 2013: 57). To that extent, the EEZ does in effect act as maritime prolongation of the coastal state's landed property. Yet, the technical and operational complexity involved in prospecting and exploiting offshore oil and gas, for example, places high barriers to entry which only powerful states (through National Oil Companies) and multinational corporations can afford to meet. Thus relative newcomers to offshore hydrocarbons like Ghana and Equatorial Guinea rely overwhelmingly on foreign companies to deliver the costly infrastructure required for exploring, drilling, extracting and transporting deep-water crude and gas (Chalfin, 2015). The resource flows at sea are replicated in the mobility of both maritime installations (in the form of mobile deep-water drill rigs, or floating production storage and offloading vessels – FPSOs) as well as in the rotating multinational workforce. The materiality of the sea here once again reproduces a terraqueous territoriality: not only is the exploitation of marine resources reliant on land-side infrastructure and property regimes (that much is fairly obvious), but the forms of surplus appropriation adopted by terrestrial sovereign states and capitalist firms are strongly conditioned by the socio-natural cycles and forces at sea.

These challenges to, and distinctive if unstable resolutions of, sovereign principles are also reflected in the existing practices of flagging at sea (see DeSombre, 2006). Designed in the 1920s to avoid US labour law and Prohibition, the legal innovation of the modern FOC originated in Panama, where formal sovereignty granted both legal and illicit American firms the right to register vessels under its flag, in return for a small fee. From the 1970s, revenue-poor countries countries like Liberia, Republic of the Marshall Islands and the Bahamas joined Panama as leading FOC states, leasing their sovereignty through the open registry system. As with the EEZ, the relationship between sovereignty, territory and appropriation for FOC boats is full of nuance and complexity. It is helpful to conceptualise FOC vessels as terraqueous territories in two senses: as sovereign spaces and as a strategy of accumulation. On the high seas, the notion of comity (involving mutual deference or courtesy between sovereigns) operates as the dominant principle when sharing a free space and settling disputes across different maritime jurisdictions. In this way, the law of the flag state establishes borders and territorialises space on board ships even when steaming through another state's sovereign waters. In such cases, the particular relation between land and sea becomes sharply apparent, as legal principles like comity articulate sovereignty, territory and appropriation in a global ocean otherwise deemed to be lawless and unruly. Moreover, given that a characteristic of the open registry is the ability of shipowners to 'buy' a sovereign and thus the legal jurisdiction that regulates their activities, shipowners produce territory as an accumulation strategy. Shipowners use sovereignty invested in state jurisdiction to cut crew costs and undermine the self-organisation of labour, as well as minimising tax bills and avoiding agreements on fishing quotas.

The cruel irony is of course that the system of 'open' registers disguises some of the world economy's worst working practices and most opaque ownership and taxation structures. The FOC regime guarantees all the surface speed, flexibility and mobility privileged by capital while condemning those who work and live in the ship's lower quarters to confinement, regimentation and domination. In contrast to factories and fields, fishing circumscribes physically the labour process to floating platforms of production that can transcend jurisdictions in various ways (e.g. legally through FOC and/ or geographically following the fish between EEZs and the legal grey zone of the high seas). The ship in this regard becomes what Jonathan Bach (via Bruno Latour) calls an 'immutable mobile' – an object that moves through space without thereby losing its property as a site of production (Bach, 2011). As in other sectors highly reliant on migrant labour, the shipping and fishing industries exploit the flexible, low-cost but highly controlled labour process afforded by the open registry system. Yet the difference is that, at sea, it is the floating capital that is in constant movement. Labour remains relatively static within the factory ship, and the possibilities of shore leave are highly restricted. In extreme, though hardly rare, cases, seafarers are in effect imprisoned for years on ships (or stranded undocumented on foreign ports), acting as bonded and even slave labour tied by land-side debts and obligations to shipowners and operators. The integrated network of legal-bureaucratic and market power sustaining the open registry regime from land thus contrasts – and has a corollary – with the isolation, precariousness and vulnerability of fishers and seafarers working on FOC ships at sea. These uniquely terraqueous organisations of space deliver distinctive geographies of labour exploitation, identity and solidarity mentioned above. In sum, the 'open registry' regime illustrates how fishing vessels in particular are never far off land when they're at sea: they carry with them all of the characteristics of a land-based labour process associated with, say, mining – ethnic segmentation of the workforce, strict labour discipline, repetitive tasks, combination of workplace and lifeworld – in a single confined space. Most important for our purposes, the FOC is a vector for the inherently multinational organisation of maritime logistics where at any given time a 'Greek owned vessel, built in Korea, may be chartered to a Danish operator, who employs Filipino seafarers via a Cypriot crewing agent, is registered in Panama, insured in the United Kingdom, and transports German-made cargo in the name of a Swiss freight forwarder from a Dutch port to Argentina, through terminals whose concessions have been granted to port operators from Singapore' (Bottalico, 2018: 44).

Towards a terraqueous International Relations

The centrality of the social system we call capitalism in shaping the relationship between international relations and the sea has been implicit throughout this chapter. Plainly, the premise throughout has been that this unique mode of production conditioned both the form and content of a maritime factor in the development of modern international relations. The socio-economic, technological, juridical and political innovations spurred on by the ceaseless quest for value, characteristic of industrial capitalism in particular, have now encompassed almost the entirety of the planet, as illustrated above. Yet IR has been attentive only to a fraction of these transformations, remaining stubbornly terrestrial in its core assumptions about the nature of world politics. Even among those theorists of the sea like Alfred Thayer Mahan or Sir Julian Corbett, who might reasonably be incorporated into the canon of classical IR, the ocean world remains an unchanging geopolitical fact: a highway of commerce and communication to be commanded and protected from enemy incursion or control (Stafford Corbett, 1911; Mahan, 2005).

The argument presented here, on the other hand, emphasises the dynamic nature of the sea, and indeed that of its relationship with land. Considering the uneven and protracted interaction between these geophysical entities, I have suggested, brings to the fore multiple dimensions of international relations that might otherwise be submerged or entirely forgotten in mainstream accounts of world politics. Among the former are the maritime origins of many key practices of modern international relations, including global governance, international law or multilateral cooperation. Evidently, these norms and institutions did not emerge, Neptune-like, fully formed from the sea, but they are the expression of several specifically oceanic features identified throughout this chapter which are inescapable from socio-economic and political developments on land. In return, terrestrial institutions and practices have clearly also transformed the nature of the sea – most obviously through anthropogenic climate change. But it is perhaps the analytical neglect of the many seaborne experiences of international relations that is most damaging for our discipline. For in marginalising or overlooking the profoundly transnational lives spent and sacrificed at sea, there is a real danger we continue to reify terra-centric accounts of world politics focused around the dominant notions of sovereignty, war, diplomacy and trade. The sea offers us a vantage point of international relations which immediately and almost inherently relativises the anchoring of world politics on land.

This, to conclude, is not simply a plea to 'bring the sea back in' to IR (though obviously that is part of the task). It is mainly an argument for a genuinely amphibious or terraqueous IR that takes seriously the material properties of land and sea in their complex interaction. Doing so can begin to remedy IR's 'seasickness' and deliver a much more complete account of the geopolitical nature of world politics, while simultaneously recognising the variable (though not 'stadial') historical temporalities being operationalised both within human societies, and in our collective relation to the non-human world. As has been suggested throughout this chapter, we can thereby account for the more prominent dimensions of contemporary interstate relations as part-manifestation of the more mundane, transnational experiences of people at, through or near the sea. A terraqueous IR might in this way both contribute to a spatio-temporal de-centring of the modern international system while still acknowledging that it is the material effects of capitalism as a historically peculiar way of organising the social transformation of nature which most profoundly conditions our planetary present and future.

Notes

1 This chapter draws on a book written with Liam Campling, *Capitalism and the Sea: The Maritime Factor in the Making of the Modern World* (London: Verso, 2021) and a jointly authored article 'Capitalism and the sea: sovereignty, territory, and appropriation in the global ocean', *Environment and Planning D: Society and Space* 36(4) (2018): 776–94.
2 For an iconic case, see Harcourt (2006).
3 See Balachandran (2012); for the eighteenth century, see the classic study by Linebaugh and Rediker (2000).

References

Alimahomed-Wilson, J., and I. Ness, eds (2018). *Choke Points: Logistics Workers Disrupting the Global Supply Chain*, London: Pluto.

Amrith, S. (2019). 'Pollution in India could reshape monsoons', *The Atlantic*, 21 January, www.theatlantic.com/science/archive/2019/01/indias-monsoon-powerful-agent-climate-change/579940/ (accessed 2 December 2021).

Anievas, A., and K. Niçancloğlu (2015). *How the West Came to Rule: The Geopolitical Origins of Capitalism*, London: Pluto.

Bach, J. (2011). 'Modernity and the urban imagination in economic zones', *Theory, Culture & Society* 28(5): 98–122.

Balachandran, G. (2012). *Globalizing Labour? Indian Seafarers and World Shipping, c. 1870–1945*, New Delhi: Oxford University Press.

Benton, L. (2010). *A Search for Sovereignty: Law and Geography in European Empires 1400–1900*, Cambridge: Cambridge University Press.

Bonacich, E., and J. B. Wilson (2008). *Getting the Goods: Ports, Labor, and the Logistics Revolution*, Ithaca, NY: Cornell University Press.

Bottalico, A. (2018). 'Across the chain: labor and conflicts in the European maritime logistics sector', in J. Alimahomed-Wilson and I. Ness, eds, *Choke Points: Logistics Workers Disrupting the Global Supply Chain*, London: Pluto, 35–49.

Brandon, P. (2015). *War, Capital, and the Dutch State (1588–1795)*, Leiden: Brill.

Brathwaite, K. (1974). *Contradictory Omens: Cultural Diversity and Integration in the Caribbean*, Mona: Savacou Publications.

Bridge, G. (2013). 'Territory, now in 3D!', *Political Geography* 34(A1–A4): 55–7.

Broeze, F. (1981). 'The muscles of empire: Indian seamen and the Raj, 1919–1939', *Indian Economic and Social History Review* 1: 43–67.

Broeze, F. (1991). 'Militancy and pragmatism: an international perspective on maritime labour, 1870–1914', *International Review of Social History* 36(2): 165–200.

Cederlöf, G. (2013). *Founding an Empire on India's North-Eastern Frontiers, 1790–1840*, Oxford: Oxford University Press.

Chalfin, B. (2015). 'Governing offshore oil: mapping maritime political space in Ghana and the Western Gulf of Guinea', *South Atlantic Quarterly* 114(1): 101–18.

Corbin, A. (1994). *The Lure of the Sea: The Discovery of the Seaside in the Western World 1750–1840*, trans. Jocelyn Phelps, Cambridge: Polity.

Couper, A., H. D. Smith and B. Ciceri (2015). *Fishers and Plunderers: Theft, Slavery and Violence at Sea*, London: Pluto.

Cowen, D. (2014). *The Deadly Life of Logistics: Mapping Violence in Global Trade*, Minneapolis, MN: University of Minnesota Press.

Creighton, M. S., and L. Norling, eds (1996). *Iron Men, Wooden Women: Gender and Seafaring in the Atlantic World, 1700–1920*, Baltimore, MD: Johns Hopkins University Press.

DeSombre, E. R. (2006). *Flagging Standards: Globalization and Environmental, Safety and Labor Regulations at Sea*, Cambridge, MA: MIT Press.

DeSombre, E. R. (2011). *Fish*, Cambridge: Polity.

Featherstone, D. (2015). 'Maritime labour and subaltern geographies of internationalism: black internationalist seafarers' organising in the interwar period', *Political Geography* 49: 7–16.

Fink, L. (2014). *Sweatshops at Sea: Merchant Seamen in the World's First Globalized Industry, from 1812 to the Present*, Chapel Hill, NC: University of North Carolina Press.

George, R. (2013). *Deep Sea and Foreign Going: Inside Shipping, the Invisible Industry That Brings You 90% of Everything*, London: Portobello Books.

Haas, P. M. (1990). *Saving the Mediterranean: The Politics of International Environmental Cooperation, Political Economy of International Change*, New York: Columbia University Press.

Hannigan, J. (2016). *The Geopolitics of Deep Oceans*, Cambridge: Polity Press.

Harcourt, F. (2006). *Flagships of Imperialism: The P&O Company and the Politics of Empire from its Origins to 1867*, Manchester: Manchester University Press.

Haushofer, K. (1938). *Geopolitik des Pazifischen Ozeans: Studien uber die Wechselbeziehungen zwischen Geographie und Geschichte*, Heidelberg and Berlin: Kurt Vowinckel Verlag.

Klose, A. (2015). *The Container Principle: How a Box Changes the Way We Think*, Cambridge, MA: MIT Press.

Lambert, A. (2018). *Seapower States: Maritime Culture, Continental Empires and the Conflict that made the Modern World*, New Haven, CT and London: Yale University Press.

Levinson, M. (2006). *The Box: How the Shipping Container Made the World Smaller and the World Economy Bigger*, Princeton, NJ and Oxford: Princeton University Press.

Lewis, D. (1994). *We, the Navigators: The Ancient Art of Landfinding in the Pacific*, Honolulu, HI: University of Hawai'i Press.

Linebaugh, P., and M. Rediker (2000). *The Many-Headed Hydra: The Hidden History of the Revolutionary Atlantic*, London: Verso.

Macfarlane, R. (2011). *The Old Ways: A Journey on Foot*, London: Hamish Hamilton.

Mahan, A. T. (2005 [1890]). *The Influence of Sea Power upon History, 1660–1783*, Cambridge: Cambridge University Press.

Malm, A. (2016). *Fossil Capital: The Rise of Steam Power and the Roots of Global Warming*, London and New York: Verso.

Mearsheimer, J. (2014). *The Tragedy of Great Power Politics*, New York: W.W. Norton.

Moore, J. W. (2015). *Capitalism and the Web of Life: Ecology and the Accumulation of Capital*, London: Verso.

Mulich, J., with L. Benton (2015). 'The space between empires: coastal and insular microregions in the early nineteenth-century world', in P. Stock, ed., *The Uses of Space in Early Modern History*, London: Palgrave Macmillan, 151–71.

Müller, S, M., and H. J. S. Tworek (2015). '"The telegraph and the bank": on the interdependence of global communications and capitalism, 1866–1914', *Journal of Global History* 10: 259–83.

Murphy, C. N. (1994). *International Organization and Industrial Change: Global Governance Since 1850*, Cambridge: Polity Press.

Paine, L. (2015). *The Sea and Civilization: A Maritime History of the World*, New York: Atlantic Books.

Phillips, A., and J. C. Sharman (2015). *International Order in Diversity: War, Trade and Rule in the Indian Ocean*, Cambridge: Cambridge University Press.

Prange, S. R. (2018). *Monsoon Islam: Trade and Faith in the Medieval Malabar Coast*, Cambridge: Cambridge University Press.

Rediker, M. (1987). *Between the Devil and the Deep Blue Sea: Merchant Seamen, Pirates, and the Anglo-American Maritime World, 1700–1750*, Cambridge: Cambridge University Press.

Risso, P. (1995). *Merchants and Faith: Muslim Commerce and Culture in the Indian Ocean*, Boulder, CO: Westview Press.

Sampson, H. (2013). *International Seafarers and Transnationalism in the Twenty-First Century*, Manchester: Manchester University Press.

Schmitt, C. (1997 [1942]). *Land and Sea*, Washington, DC: Plutarch Press.

Sheriff, A. (2010). *Dhow Cultures of the Indian Ocean: Cosmopolitanism, Commerce and Islam*, London: Hurst and Company.

Shilliam, R. (2015). *The Black Pacific: Anti-Colonial Struggles and Oceanic Connections*, London: Bloomsbury.

Spykman, N. (1944). *The Geography of the Peace*, New York, Harcourt, Brace and Company.

Stafford Corbett, J. (1911). *Some Principles of Maritime Strategy*, New York: Pantianos Classics.

Steinberg, P. E. (2001). *The Social Construction of the Ocean*, Cambridge: Cambridge University Press.

Steinberg, P., and K. Peters (2015). 'Wet ontologies, fluid spaces: giving depth to volume through oceanic thinking', *Environment and Planning D* 33(2): 247–64.

Stopford, M. (2009). *Maritime Economics*, 3rd edn, London: Routledge.

Subramanian, L. (1999). *Medieval Seafarers*, New Delhi: The Lotus Collection.

Verne, J. (2014 [1870]). *20,000 Leagues Under the Sea*, London: Bloomsbury.

Walcott, D. (1978). 'The Sea Is History', in E. Baugh, *Derek Walcott: Selected Poems*. New York: Farrar, Straus and Giroux, 137–9.

Weiss, H. (1996). *Solidarity – The First 100 Years of the International Transport Workers' Federation*, London: Pluto.

Weiss, H. (2017). 'The international of seamen and harbour workers: a radical global labour union of the waterfront or a subversive world-wide web?' in H. Weiss, ed., *International Communism and Transnational Solidarity: Radical Networks, Mass Movements and Global Politics, 1919–1939*, Leiden, Brill, 256–317.

Young, O. R. (1992). *Arctic Politics Conflict and Cooperation in the Circumpolar North*, Chicago: Chicago University Press.

2

The symbolic space of the sea: mythologising a nation, performing an alliance

Maria Mälksoo

What is Estonia? How does deterrence work? Why do they both need the sea to be? This chapter offers an empirically Baltic Sea-bound exploration of the sea as an important symbolic realm in world politics for the construction of national mythologies and the practical performance of alliances. Ever since Homer's *The Odyssey*, the sea is frequently depicted as a liminal space, and the journey across the sea an important rite of passage in fictional and political imaginations alike. Despite its well-acknowledged economic, political, strategic and affective value, the sea as the subject of political symbolisation and a source of symbolic power in world politics has received limited scholarly attention (see Klein and Mackenthun, 2004). As any material object chosen to represent something beyond its immediate self, 'something immaterial or abstract', such as a 'being, idea, quality, or condition' that has 'sacred' importance,[1] the sea has multiple symbolic meanings and functions. Yet, its politically framed symbolic significance remains poorly understood and socially thin: the discipline of International Relations (IR) tends to regard the sea either as an economic resource (a space to be controlled), or a strategic playground (a space to be tamed) to better travel and transport goods to the humanly more habitable land (a space to be traversed) (de Carvalho and Leira, 2022; this volume). For the field of international politics, the sea has historically been the subject of physical and mental colonisation, exploitation, extraction and governance. Lost is the emotive power of the sea as a symbolic resource and its cultivation as a political symbol.

This chapter turns to social theory and anthropology to elucidate the sea as a symbolically loaded space with important political
meanings and functions in the present. It does so through deepening
the IR attention on the Baltic Sea as a well-trodden path in the study
of regionalisation (e.g. Joenniemi, 1993) and ecological governance
(Tynkkynen, 2013, 2015) via two illustrative cuts to expound the
theoretical argument. First, a reading of *Silver White* (1976) – a
historical travelogue theorising the Baltic-Nordic connectivities
(and beyond) through the ancient seafaring history by an Estonian
writer, ethnographer, film-maker and the first post-1991 president
Lennart Meri (1929–2006)[2] as an instance of the purposive construction of the sea as a national mythscape; and second, the emerging maritime posture and posturing of NATO in the Baltic Sea
region as a case of ritualised performance of deterrence towards
Russia. In both instances of cultivating a national mythscape via the
sea and performing a multinational military alliance via exercising
extended maritime deterrence, the Baltic Sea emerges as a crucial
arena for creating and enacting political subjectivities and communities in world politics.

The chapter unfolds in four stages. I start by conceptualising the
sea as a symbolic space, susceptible to be framed as a mythscape
(Bell, 2003), and further prone to ritualisation as a culturally strategic way of acting in the world and exercising power (Bell, 2009).
A brief methodological explication and the two empirical snapshots
follow, illuminating the mythologised and ritualised features of the
historic and contemporary Baltic Sea space in distinct ways. The
conclusion weaves the main threads of the chapter together.

Seascape as a symbolic space

Thinking with the editors' threefold typology of the sea as a
space in world politics, the sea has figured either as an obstacle
to be mastered, a space to be traversed or a set of resources to be
extracted and controlled in global politics (de Carvalho and Leira,
this volume). Yet, besides its physical and social dimension, the sea
is also a symbolic terrain – a rich realm for animating historical,
poetic and affective imaginations; a space which keeps generating
new spaces. According to Sherry B. Ortner (1973: 1340), some

symbols summarise a variety of human experiences, while others help to make sense of these experiences and locate them in cultural categories. The sea embodies both possibilities, allowing to link the manifold individual human experiences with the cultural dramatisation of a certain collective biographical narrative, identity and value system condensed in this medium of human interaction. The symbolic form of the sea hence allows structuring of the production of both meaning and experience (see Cassirer, 1946: 8; Bartelson, 2014: 15). Such a twofold potential is captured in Abner Cohen's discussion of the ambiguity and multidimensionality of symbols – as effectively bifocal phenomena, focusing on existential ends on the one hand and political ones on the other (Bell, 2009: 184).

The physically evocative form of the sea – its perpetual flow and seeming unboundedness to the human eye, along with its historical role in connecting people – has contributed to the fecundity of invoking the sea as a mythical backdrop for nation- and region-building. Fernand Braudel (1995) introduced the concept of the sea as the facilitator of interchange between regions located around its coastlines. The idea of the sea as the catalyst and enabler of international contact is underpinned by the fact that during the early modern period, transport over water was easier and cheaper than over land: hence Braudel's suggestion that regions connected by waterways interacted more than regions only connected by land (van Bochove, 2007: 155). This is also reflected in Derek Walcott's (1978) poetic metaphor of the sea as 'history', evoking the sea as primarily a social and political contact zone – 'a meeting place of different cultures', and as such, 'a complex and polysemic dialogue' (Klein and Mackenthun, 2004: 1–2). As a space, the ocean is very much part of history and within history, regardless of the romanticised and mythical accounts often reducing the sea to a symbol of unruliness or irrationality outside and beyond history (Klein and Mackenthun, 2004: 1–2). As with the sacred, the sea is capable of evoking the feeling of terror, 'the awe-inspiring mystery (*mysterium tremendum*), the majesty (*majestas*) that emanates an overwhelming superiority of power' (Eliade, 1959: 9, citing Rudolf Otto's *Das Heilige*, 1917). Historically, the waters have been religiously valorised:

> The waters symbolize the universal sum of virtualities; they are fons et origo, 'spring and origin', the reservoir of all the possibilities

of existence; they precede every form and support every creation. One of the paradigmatic images of creation is the island that suddenly manifests itself in the midst of the waves. On the other hand, immersion in water signifies regression to the preformal, reincorporation into the undifferentiated mode of pre-existence. Emersion repeats the cosmogonic act of formal manifestation; immersion is equivalent to a dissolution of forms. This is why the symbolism of the waters implies both death and rebirth. (Eliade, 1959: 130)

Due to its historically, culturally and emotionally stimulating power, the seaward domain is also a crucial symbolic resource to imagine, glorify and mobilise visions of the past for the purposes of the present. The ocean is a 'nostalgic space' (Steinberg, 1999b: 417), yet a deeply social one, constructed amid competing interests and priorities and transformed along with social change (Steinberg, 2001: 207). The sea has provided 'a foundational, if somewhat ephemeral, grounding for some of the major geopolitical, geocultural, and geoeconomic referents of our time, including the North Atlantic Treaty Organization' (Steinberg, 1999c: 370).

As a starting point for unfolding the political symbolism of the sea for constructing historical imaginaries of nations, I take my cue from Duncan Bell's notion of a mythscape, understood as 'the perpetually mutating repository for the representation of the past for the purposes of the present' (Bell, 2003: 66). Meri's *Silver White* serves as an evocative example of a consciously recited mythscape where the narrator weaves together threads from his intricate reconstruction of the ancient sea-bound historiography of the Baltic Sea region. The purpose of such an allegorical endeavour is not to sacralise the sea per se, but to articulate an inspiring message for the present: to effectively gather up courage for the spiritual survival of the fast-stagnating Soviet Estonia in the mid-1970s when Meri's book was first published. The sea, and the history it entails, becomes a deliberately hailed source of identity strength: an important anchor for reminding a tiny nation supposedly on the verge of political extinction of the span and multi-vectored interwovenness of its past via highlighting Estonians' historical connectivities with many other nations through the sea. The openness of the past provides a stark contrast to the artificial boundaries of the Soviet present, as *inter alia* illustrated by the closing off of the access to the Baltic Sea for the local peoples during the Soviet era in Estonia (1944–91).

The historical reminders of the mutuality of past influences, and the occasional east–west vector of these influences, enables Meri, the mythologist, to make the small nation bigger at least in imagination. *Pace* Meri, for Estonia as 'a narrow strip of coastland, almost a peninsula, surrounded by the sea on two sides … the role of the sea in the evolution of our culture has been somewhat more significant than presumed thus far' (Meri, 2008: 538). The seaways are the 'axis of our life' and the sea is 'too active an economic, political, psychological and emotional factor to silently disappear from the life of a coastal nation' (Meri, 2008: 533–4). Hence his rationale of making sense of this bundle of connectivities head-on in order to draw strength for the dispiriting political present of the slowly extinguishing Estonian nation at the height of the Sovietisation policy. The sea as the outward facing pillar of the Estonian identity is thereby juxtaposed to the traditionally introspective Finno-Ugric life-world and the related mythologies focused on the forest (i.e. *metsaeestlus* or the so-called 'wood Estonianness'; Mikita, 2008, 2013, 2015, 2017; Kant, 1935; Loorits, 2000).

But the sea can also be a site of concretely defined public rituals (including political, military and security rites) (see Etzioni, 2000). The shipboard culture is a deeply ritualised phenomenon and as such soundly studied by anthropologists and culturologists (e.g. Dening, 1992; Steinberg, 1999a: 261–2; Weissman, 2015). Navies have a distinct identity and a set of rituals as seamen, i.e. people living on the sea. Meanwhile, the sea as a rich canvas for the strategic use of its symbolic features has received more modest scholarly attention, all the less so in IR. Catherine Bell's framework of ritualisation, defined as 'a specific embodiment and exercise of power' which endows social agents 'with some degree of ritual mastery', offers a useful lens here (Bell, 2009: 179, 141). Employing the seascape as a source material for national mythology can act both as a medium of ritualisation and an end product of it.

For Bell, ritual practices are created with 'an intent to order, rectify, or transform a particular situation', albeit 'ritualised agents do not know what what they are doing does' (Bell, 2009: 108). Ritual can hence be a strategic way to construct a type of tradition (to 'traditionalise'), but 'in doing so it can also challenge and renegotiate the very basis of tradition to the point of upending much of what had been seen as fixed previously or by other groups' (Bell, 2009: 124).

NATO's post-2014 performance of maritime deterrence in the Baltic Sea region illustrates how the sea can be utilised as a stage for ritualised politico-military practices, such as extended deterrence. A theatrical dramatisation of maritime deterrence serves as a stopgap of the practical deficiencies of NATO's current defence and deterrence posture in its north-eastern flank. The access of NATO's military reinforcement forces via the sea to the politically and militarily exposed Baltic states is vital for the credibility of the Alliance's deterrence in the region vis-à-vis Russia, as NATO's land and air access to the Baltic states remain potentially hampered by the vulnerability of the so-called Suwałki Corridor due to Russia's anti-access/area-denial (A2/AD) capabilities.

Methodological considerations

Weaving together a historico-political vision of an authoritative political individual and a strategic practice of the arguably most powerful modern military alliance might strike a worn-out chord of attempting thus to compare apples and oranges. I have selected these two empirical illustrations for three reasons: first, because they both illuminate a strategic use of maritime space while capitalising on the political symbolism of the sea; second, because the Baltic sea-space figures prominently in both cases; and third, because they allow the empirically pinpointing of the construction of the sea as a symbolic space and its significance for distinct political subjectivities (i.e. a nation and alliance/security community).

I rely on both primary and secondary sources in this enquiry of the symbolic potential and ritualisation of the sea. In the case of Meri's *Silver White*, I offer a close reading of the text itself, along with an engagement of pertinent secondary literature and some linking of the author's early musings on the sea with his later presidential public discourse. The exploration of NATO's emergent maritime deterrence posture in the Baltic Sea relies on the publicly available data and images of the recent prominent naval exercises in the region via NATO's own official sites and various think tank reports. In fleshing out the argument on the symbolic meanings and uses of the sea, the analytical focus is on the definition of political boundaries through the mythologisation and ritualisation of the sea-space.

Mythologising a nation through the sea

Lennart Meri's *Silver White* (2008 [1976]) is a semi-popular, semi-scientific historical fantasy about the interconnected pasts of ancient Estonians and other Baltic Finnic people. Setting out to explore 'which ships have sailed on our seas? why have we forgotten them?' (Meri, 2008: 203), Meri's book offers a bold reconstruction of ancient Estonians' past through maritime connectivities. *Silver White* reads as a mythologised re-enactment of a multi-threaded story of origin for a small nation which was denied an autonomous political present and future in the deadly embrace of the Soviet empire at the time of the volume's original publication in the mid-1970s.

A central theme in *Silver White* is the sea as the principal structuring agent of the Estonian culture, the nature and worldview of the local people. In Meri's account, the sea acquires a strongly symbolic significance, beyond the traditional breakdown of human–marine interactions which would regard the sea as a resource provider, transport surface, and a battleground or 'force-field' (Steinberg, 2001: 11) or a space of control and extraction, a space to be traversed, a space to be tamed (de Carvalho and Leira, this volume). In *Silver White*, the Baltic Sea figures emphatically as 'a great artery of communication' (see Spykman, 1944: 35). The book evocatively subscribes to Philip Steinberg's take on the sea as not 'a neutral surface across and within which states have vied for power and moved troops' but a socially constructed space throughout history, much alike the nation states themselves (Steinberg, 2001: 18). Identity is acquired via and through the sea (it 'comes from the sea'); the nation-building and reclaiming of history works through the emphasis on historic seafaring connectivities. As a purposive mythologising endeavour, *Silver White* is most unequivocally an attempt to return (and introduce) a *longue durée* history to the people who were being actively assimilated into the Soviet geopolitical and cultural formation at the time of the book's origination. The traversable quality of the sea enables a wandering politico-historical imagination where the space of the sea is sought to be *untamed* for the sake of temporally mastering the past as a solace in the politically poorly controllable present.

Meri defined the genre of his book[3] as a travelogue himself. It is a very special kind of a travelogue, however – more of a visionary historical mind-travel where the author's free-flowing and often audacious associations, drawing on his education in history and the many expeditions to the Baltic Finnic peoples, touch base with and weave together insights from folklore, astronomy, geology, linguistics and navigation along with the writings of historical and contemporary seafarers, merchants, and contemporary scholarship marshalled from all the aforementioned fields. Most notably, *Silver White* is an intellectual travelogue, interweaving historically distinct travelogues touching upon the Baltic Sea region by Meri's passionate quest to connect the sea-space surrounded by the Baltic coasts with the important historical explorers and waterways of the world. While the book can be read as a fact-based attempted reconstruction of the ancient history of the Baltic Sea lands, its authorial position is unashamedly interpretive, roaming wildly with building unorthodox historical connections and associations in the tradition of 'it could have been this way – and perhaps even more'. The question that the author frequently posits to himself throughout the book – namely, 'Where do the facts end, where does poetry begin?' – characterises Meri's oeuvre equally poignantly (Meri, 2008: 288). For Meri, poetic texts are useful sources of history as a factual thread supposedly still runs through them, leading us to 'a reasonable and palpable piece of information' (Meri, 2008: 509). Accordingly, 'history-writing starts where poetry ends', but '[s]ometimes they brush against each other' (Meri, 2008: 539). *Silver White* is, by Meri's account, 'a bridge between poetry and sciences' (Meri, 2008: 254). Accordingly, its methodical historical reconstruction is sprinkled with poetic vignettes throughout (e.g. 'Sail. Sail. Sail. The whole world.'; Meri, 2008: 393). For Meri, the prosaism of geography and the iconography of maps needs occasional poetry to complete the picture of imaging the past and historicising the Baltic Sea (Meri, 2008: 228).

The overarching aim behind this chronicle of the *Silver White* waterway (i.e. the Way East or the *Austrvegr* by its Scandinavian name)[4] is an endeavour to substantiate (and interpellate the Estonians to) a self-imagination as a very old nation that has historically been thoroughly connected to the peoples both in the North, the West and the East. Meri's *Silver White* effectively seeks to carve out an autobiography of the Estonian nation through the connectivities with many other, and often distant, cultures. In social interactionist terms, the book is spearheaded at unpacking the historical function of the Baltic Sea in facilitating the Estonians' Me-identity, their social sense of self (see Mead, 2015). Mythologising the ancient seafaring contacts of Estonia concurrently stakes a claim for the country's connection to Europe. The book seeks to reanimate the lived experience of the Baltic Sea in the collective memory (as sedimented in the folklore, myths and language of the Baltic Finnic peoples) and the historiography of ancient travellers, such as the Greek Pytheas of Massalia (around 350 BC) and Arabian al-Idris. According to Meri's unorthodox hypothesis, Ultima Thule, an island at the edge of the world as described by Pytheas, was likely the largest Estonian island Saaremaa (literally, the isle-land); and a city depicted in the Arabic traveller al-Idris's travelogue turns out to be Tallinn – the largest city and the historical capital of Estonia, located at the Baltic Sea, 80 km off the Finnish coast. Placing Pytheas's Thule to Saaremaa (unlike the majority of existing accounts which have located it either to Shetland, Iceland, or the Norwegian coast), Meri's unconventional proposition has generally been interpreted as an inspired, yet factually overstretched, fantasy by other historians (e.g. Kõiv, 2017). But Lennart Meri was not a scholar as much as he was a visionary. For Meri, paraphrasing his own words, 'metaphors, myths and facts were the building blocks of equal weight in his work' (Meri, 2008: 136). *Silver White* is first and foremost an active identity-building project, taking the Sartrean proverb 'we are what we make of what others have made of us' at its word.

Meri's daring hypothesis of locating the Antique Thule to Saaremaa remains intriguing for its ambition to give to the supposedly politically futureless Estonian nation the courage to dream about having mattered in the ancient past at least. As such, his theory has found ample social resonance among the Estonians and acquired myth-like connotations. *Silver White* is underpinned by Meri's belief that the catastrophe of Kaali[5] left a significant imprint 'in the minds, languages and a sense of the world to all nations around the Baltic Sea' (Meri, 2008: 528). The wildly imaginative connecting of an obscure Baltic island to an enigmatic antique seafarer's text also has a reassuring message against the backdrop of the political reality of the time: what happened in the ancient Estonian land mattered to people and places far beyond it.

The Baltic Sea accordingly emerges as the central enabler, a communicator and a connector in Meri's account: it is mythologised as a socially thick space in the past and presented as a source of symbolic strength for the peoples along its coasts in the present. Through their strategically important location at the Baltic Sea coast, the space known today as Estonia emerges as a critical connection between the West and the East in Meri's account (Meri, 2008: 363). Cultural contacts, *pace* Meri, 'have always taken place, as far as the eye can reach … The pathos of evolution and discoveries lies in the observation of cultural contacts and their conscious development' (Meri, 2008: 538). The sea thus becomes a space for reimagining borders, in order to mentally break away from the political suffocation of the 1970s marked by the political stagnation in the U.S.S.R., the intensifying Russification of the Baltic republics, and the Estonian society's emotional resignation by the persuasion of imaginative history.[6] With *Silver White*, Lennart Meri is metaphorically attempting to do what the popular Estonian folk singers The Johanson Brothers envisioned in their song 'Lahtilükkamine' at the time of the Singing Revolution in Estonia (1988): namely, detaching the whole country from its eastern neighbour by giving it a collective push to turn Estonia into a floating island and send it off on a voyage along the world seas. The immediate political impetus to break free from the crumbling Soviet empire became part of a broader impulse to 'take a tour around the world … through the Danish straits … to the world seas … to find another place for our soil' in this song (Jaak Johanson, lyrics and melody). The gist of such an imaginary push is palpable in Meri's *Silver White* musings already a decade earlier. His spatial imaginary is markedly waterbound rather than land-centric: Europe as a whole appears as 'a tiny peninsula' in his account; 'a seed of the burdock at the tail of a huge continent' (Meri, 2008: 295).

Meri's passionate co-travelling with historic seafarers, linguists and historians is animated by a desire to discover, but importantly also to recognise the familiar in the distant references made about the land known today as Estonia in various scholarly and more popular texts. *Silver White* resonates with the observation Jorge Carrión makes in his inimitable travelogue of bookstores across the world, namely that '[w]e travel to discover, but also to recognize' (Carrión, 2013: 113). To *Silver White* (as to any identity-building project), we might add: but also, *to be recognised*. The book struggles against the tide of a commonly gloomy collective self-conception of 'small peoples' – the 'lack of a sense of an eternal past and future' (see Abulof, 2009: 235). To give hope for a future to the speedily Sovietised Estonians at the time, Meri proceeded from giving his fellow Estonians first a past and the reasons to draw pride from it. As such, *Silver White* is also an emphatic exercise in symbolic power: 'the power to make things with words … the power to conse-crate or to reveal things that are already there' (Bourdieu, 1989: 23).

Meri's historicisation of the Baltic sea-space is concurrently its mythologisation. This strategic move entails the reimagination of ancient seafaring connectivities from an experiential perspective by historicising out of the earliest available seafarers' stories[7] along with the emphatic myth-cultivation of a metalanguage for crossing cultures in space and time (see Dening, 2004: 15). *Silver White* is an illustration of the truism that myths 'do not simply evolve unguided, without active agency' (Bell, 2003: 75). Instead, they are consciously constructed and shaped. *Silver White* reads in the first place as an intentionally created mythscape for an Estonian nation – 'a discursive realm, constituted by and through temporal and spatial dimensions, in which the myths of the nation are forged, transmitted, reconstructed and negotiated constantly' (Bell, 2003) – yet one that is decidedly open to nuance and contradictions, rather than reifying a single story.

At the centre of *Silver White*'s distinctly *unlanded* mythscape is the myth of a 'very old' (Estonian) nation, cultivated by the trope of the early Baltic Finnic ancestors allegedly having set foot on the coastland of the Baltic Sea five thousand years ago (see Foucault, 1965).[8] The Baltic Finnic people reaching the Baltic space called into a life 'the sea culture and the sea cult' which developed under the strong influence of early Estonians' maritime neighbours (Meri, 2008: 19). Sea culture, in its turn, created 'a sea religion where the ship had a central place' (Meri, 2008: 254). The sea animated the local people and transmitted some of the 'Indo-European dyna-mism' to the historically forest- and field-bound people (Meri, 2008). While the positively connoted quality of dynamism links the ancient Estonian people to the influences from the Indo-Europeans in the North and the West, *Silver White*'s envisioning of the cultural contacts facilitated by the Baltic Sea is quite a leap from a simple

East–West dichotomy as a standardly imagined negative–positive pair. As Meri maintains, 'it is strange to depict cultural history as a long-distance run where someone will finish first, and someone has to accordingly remain the last... The most important element in the Northern European history is obviously not one tribe or another, but a photosynthesis' (Meri, 2008: 202–3).

In his later political career, Meri would occasionally return to the core themes of his masterpiece, such as the openness of the world, the importance of communication, curiosity and discovering of new things. The sea, seafaring and maritime connectivities continued to figure prominently as Meri's favourite metaphors in his figuratively rich and imaginative political speeches as the first president of post-1991 Estonia. Speaking at the University of London's School of Slavonic and East European Studies in 2000, President Meri introduced his country as 'an island in an ocean of Indo-European languages, an island that surfaced long before the ocean surrounding us became Indo-European' (Meri, 2000a). An island, however, is by definition more open to the outside world than a landlocked country. As an opposite of 'a mausoleum or a museum', an island is exposed to all external influences ... to the extent that the navigation season, customs, curiosity and the market allow it to be. It is this constant conflict of inclusive and exclusive qualities that allows a culture 'so small in numbers' to not drown in the Indo-European ocean, to maintain its own identity (Meri, 2000a).

For Europe as a whole, as Meri maintained in a lecture held at the University of Helsinki in 1999, its most important characteristic is 'its amazingly jagged coastline, the numerous bays, gulfs, straits, peninsulas and islands, and the balance – so rare in the world – between seclusion and contact, isolation and communication' (Meri, 1999). Accordingly, it is 'the diversity of languages, and thus also cultures [that] is the synergetic factor that has made our continent, though poor of natural resources, the global locomotive of European culture' (Meri, 1999). Due to the long coastline of the 'little peninsula of the Eurasian continent', Europe emerges as 'a model of information exchange' since 'the length of the line where the sea and the land meet has been the determining factor on which the intensity of information transfer (including skills, beliefs, goods and services) from one place to another has depended' (Meri, 2000b). The small states of Europe are accordingly deemed as the guardians of diversity, the unique particle of Europe's cultural DNA. All of this, Meri argues, has become possible because of the great connector of the sea.

Performing an alliance through maritime deterrence

The North Atlantic Treaty Organization is a maritime military alliance by name. After the end of the Cold War, NATO's original purpose as articulated by its first Secretary General Lord Ismay (to 'keep the Russians out, the Americans in, and the Germans down') gradually shifted towards a global expeditionary military role. The Alliance Maritime Strategy (2011) reflects the 'strategic drifting' away from NATO's territorial defence posture at the time of drafting the document regardless of the nominal references made to deterrence and collective defence tasks among counter-terrorism, counter-piracy and humanitarian roles on a global scale (Sundstrom, 2017).

Since the 2013–14 crisis in Ukraine, however, deterrence as a threat-based strategy of conflict management has been firmly reinstalled in NATO's vocabulary and practice. While the 1999 Washington Summit communiqué mentioned 'deterrence' just once, the 2016 Warsaw Summit statement pronounced it twenty-eight times. Deterrence has prominently returned in various practices, such as contingency and defence planning, and the explicitly ritualised performances of the Alliance, ranging from operational training, military exercises, and the deployment of multinational battlegroups in Poland, Estonia, Latvia and Lithuania. As a response to Russia's annexation of Crimea and meddling in the unfolding conflict in Donbas in 2014, NATO boosted its deterrence and defence posture in the formerly rather neglected eastern part of the Alliance, along with suspending all practical civilian and military cooperation between itself and Russia. NATO further tightened its security relationship with Finland and Sweden to buttress the situation in the Baltic Sea, and agreed on a Readiness Action Plan at the Alliance's Wales summit in 2014, envisioning a 5000-strong Very High Readiness Joint Task Force (VJTF) within the NATO Response Force. A US–European Reassurance Initiative followed course in June 2016, with an agreement to rotate an Armored

Brigade Combat Team as well as pre-positioned stocks in the Baltic states (Duke and Gebhard, 2017: 381).

The eastern flank of the Alliance has emerged as a new ritual space in the strategic debates over the major geopolitical challenges in successfully defending against Russia in the post-Crimea strategic environment, with the Fulda Gap of the Cold War years being replaced with the Suwałki Gap, a narrow land strip between Poland and Lithuania that could serve as a key transit corridor for Allied forces to deliver military aid to the Baltic states in a crisis. As the Suwałki Gap remains vulnerable to the Russian anti-access and area-denial (A2/AD) capabilities in the Kaliningrad enclave, thus threatening the access of NATO enforcements to the Baltic region by land (Frühling and Lasconjarias, 2016; Hodges et al., 2018), the maritime access of the allied reinforcements in the case of an actual crisis has gained in significance. Yet, the travelling of the necessary reinforcements as foreseen by the Readiness Action Plan adopted at the NATO Wales Summit of 2014 by air or across the Baltic Sea is equally endangered by Russian anti-air and anti-ship missiles based in the Kaliningrad enclave, for the Baltic sea-space remains still well within their range (Frühling and Lasconjarias, 2016; Kramer and Nordenman, 2016).[9] Further, although the Baltic Fleet remains the smallest of Russia's naval formations, it still is the largest naval force in the Baltic Sea. Given Russia's repeated intrusions into Swedish and Finnish territorial waters, various provocations against the Western powers[10] and the A2/AD challenge in Kaliningrad, the questions of sea control, sea denial and the amphibious landings capability have emerged as of increasing importance for the credibility of the current tripwire (and admittedly land forces-centric) deterrence model of NATO in its north-eastern flank (Kramer and Nordenman, 2016: 8; Lanoszka and Hunzeker, 2019: 114). The multiple underwater communication cables further add to the vulnerability of the Baltic Sea for the Alliance's littoral states as the severing of undersea cables to disrupt communications risks it becoming a part of a potential hybrid campaign scenario by Russia (Stavridis, 2016; see further Murphy et al., 2016: 17).

Accordingly, NATO's concern that Russia could block the Alliance's access to the Baltic Sea has helped the maritime dimension of the Alliance's re-awakened deterrent hand in the Baltic Sea region to gain prominence and pace.[11] The maritime component of NATO's Readiness Action Plan announced at the 2014 NATO Summit entailed intensified maritime patrols in the Baltic Sea (along with the Black Sea and the Mediterranean) with the Standing NATO Maritime Groups and Standing NATO Mine Countermeasures Groups, increased maritime aircraft patrols, and an expansion of the annual BALTOPS naval and amphibious exercise (see NATO, 2019).

NATO's maritime deterrence in the Nordic-Baltic space emphatically illustrates the performative nature of deterrence. This is to say that performance is not simply a way of expressing deterrence, but is itself an aspect of the very practice (see Rappaport, 1979: 176–7). The sea lends itself particularly to ceremonial spectacles of deterrence for one needs to go out of one's way to get to the sea. For Jack Goody (1961), 'ceremonial' designates a 'category of ritual' – neither religious nor magical; neither presupposing the existence of supernatural powers nor having practical ends. However, ceremonial can have certain 'ends' – the (ideological) points of view of actors – and certain 'latent functions' from the point of view of the observer (Valeri, 2014: 349). What happens at the sea needs to be communicated via special media to people. Due to the physical distance of the maritime deterrence drills from their various audiences (i.e. home and the targeted antagonist's alike), seaborne deterrence is reliant on diverse media to display the deterrent message to a sought effect. The maritime deterrence stage is hence particularly prone to ritualisation. A theatrically performative reading of deterrence highlights the co-presence of actors and audiences for the success of the ritual performance of this central security practice. A performative event of deterrence is delivered by actors (deterrers) and received by audiences (i.e. the security community participating in the act of deterring and the antagonist sought to be deterred) (Mälksoo, 2021). The audiences of maritime deterrence are usually not immediately available to witness the ritualised spectacle on the sea. Yet, the success of the performance is deemed dependent on the implicit audience of the (deterrent) message recognising the event and the related communication as convincing (Ringmar, 2012: 8).

Against this backdrop, the sea compels the mediatisation and the related aestheticisation of deterrence. The recent seaborne military exercises of NATO have clearly demonstrated the Alliance's understanding of the iconic power of photographs and videos[12] in conveying a particular narrative of the Alliance's military posture, presence and resolve in the Nordic-Baltic space. The carefully curated images and videos of NATO's maritime operations in the Baltic Sea region seek to make the meaning of deterrence 'iconically visible' for its target audiences (see Alexander, 2010: 10). These images make the idea of deterrence on the sea empirically available, relatable and 'real' for their home publics. A mediatised representation of naval deterrence rituals enables a more tangible experience of deterrence for the people (see Alexander, 2010: 11).

NATO's performance of its maritime deterrence in the Baltic Sea capitalises on the symbolism of the sea-space as a passage to land. Controlling this passage by denying it to the enemy, the sea becomes a force field in peacetime – an important space of ritualised exercises of Allied defence and cooperation for the benefit of the Alliance's vulnerable north-eastern flank, but crucially also for the Alliance's credibility and resolve towards its traditional antagonist as a whole. Performing the North Atlantic Alliance through maritime deterrence in the Baltic Sea brings the threefold thinking of the sea as an obstacle to be mastered, a space to be traversed and a pool of resources to be controlled into sharp relief along both material and symbolic dimensions.

Conclusion

The sea is prone to symbolic ascriptions. Its seeming unboundedness, flow and historical embeddedness in peoples' imaginaries of the mystical, majestic and sacred lends itself to political and cultural mythologisation and ritualisation. As a physical space and an historically laden mythscape, the sea serves as a stage for loading the world symbolically, creating affective entanglements which shape the emergence and reproduction of political subjectivities and communities of international politics in manifold ways.

Calling the attention of IR scholars to the symbolic features and uses of the sea in international identity and security politics rubs against the grain of mainstream IR, which has sought to rationalise the world, depicting the modern political reality as effectively disenchanted, and consequently denying or suppressing peoples' seemingly undying need for experiencing and practising world politics in symbolically meaningful ways. The symbolic power of the sea invites us to dip in the sea of possibilities for a better appreciation of the clout of symbols, myths and rituals in international politics at large. Thinking with the sea, as the editors of this volume call us to do, is also an invitation to think beyond the prevalent tropes of its containment and traversing. As a bountiful source of political imagination, the sea is a gift that keeps on giving – a space which generates new symbolic and ritual spaces.

Notes

1 See the *Oxford English Dictionary* definition of a symbol. For a useful introduction on core political symbols and their taxonomies, see Linklater (2019).

2 'Meri' stands for 'the sea' in Estonian. So the name of the man was eponymic to the subject he was most passionate about throughout his writings and later political speeches as a foreign minister (1990–92) and president of Estonia (1992–2001).

3 *Silver White* was first published in 1976, followed by a supplement *Hõbevalgem* (i.e. *Silver Whiter*) in 1983. The latter was deemed by the author as not a sequel, but a prequel to the original volume – indeed an introduction and the first chapter to *Silver White*. The co-edited assemblage of the two books was published in 2008, i.e. already post-humously for the author, in Estonian. The book has been translated into Finnish (1983) and Italian (2016).

4 In the Middle Ages, the Baltic Sea (German *Ostsee*, Swedish *Östersjön*, Russian *Baltiyskoye More*, Finnish *Itämeri*, Polish *Morze Bałtyckie*, Estonian *Läänemeri*) was a major highway of trade, connecting the ports of the Hanseatic League (such as Lübeck, Visby, Rostock, Gdańsk and Tallinn) (Blomkvist, 1998, 2005). The *Austrvegr* was the most important international trade route from Scandinavia to the east during the Viking Age, passing through the Eastern Baltic lands (Kiudsoo, 2016; Mägi, 2018).

5 Meri refers to the Kaali meteorite shower that hit Saaremaa estimatedly between 1530 and 1450 BC (albeit the scholars disagree about the exact timing of this rare cosmic catastrophe) (Raukas and Stankowski, 2010).

6 Twenty-five years after the original publication of the book in Estonia, Meri gave an interview wherein he offered an extended reason for writing the book in the political climate of 1970s Estonia: 'At the time, it was palpable that we were approaching an existential question: shall we persist as a nation, with a history and as carriers of certain values,

or shall we perish before the Soviet dinosaur collapses? The factor of time became of fatal significance. This is the honest explanation why I started this book then which still makes me happy today' (Meri interview with Oplatka, 2001: 208).

7 As Dening (2004: 219) reminds us: 'We never experience the past. We always historicize out of somebody else's story.'

8 Later archeological and linguistic research has offered multiple revisions and corrections to this thesis. For a good discussion of the challenges accompanying the attempt to combine linguistics and (social) archeology in the study of ethnogenesis, see Lang (2014).

9 The small width of the Baltic Sea (only 120 miles max) means that Russia's ground-to-air and ground-to-ground systems (such as the Iskander missile and S-400s based in Kaliningrad) can reach across the Baltic states, into Sweden across the Baltic Sea, covering further much of Poland (Kramer and Nordenman, 2016: 8).

10 Recent examples include the harassment of a US cargo ship on its approach to Klaipeda carrying equipment for exercise Sabre Strike 2017; the 2018 testing of missiles in Latvia's exclusive economic zone, forcing airspace closures and the re-routing of seaborne traffic; and in 2019 the trailing by Baltic Fleet corvettes of US destroyers in the Baltic Sea and the simulation of missile attacks on sea targets from Kaliningrad (Lange et al., 2019: 15).

11 NATO's Brussels Summit Declaration (2018) states: 'We are reinforcing our maritime posture and have taken concrete steps to improve our overall maritime situational awareness. We have prepared strategic assessments on the Baltic and Black Seas, the North Atlantic, and the Mediterranean. Through an enhanced exercise programme, we will reinvigorate our collective maritime warfighting skills in key areas, including anti-submarine warfare, amphibious operations, and protection of sea lines of communications. The posture will also ensure support to reinforcement by and from the sea, including the transatlantic dimension with the North Atlantic being a line of communication for strategic reinforcement' (para. 19). As NATO Secretary General Jens Stoltenberg put it: 'NATO has to be able to operate on the sea, over the sea, and also under the sea' ('The largest ever NATO Baltic exercise', 22 June 2016).

12 E.g. https://twitter.com/LTU_Army/status/1140637180907462656. See further the Twitter channel of NATO Maritime Command @NATO-MARCOM, and Forces TV, a television channel that 'focuses on all aspects of the British Armed Forces and the wider global military community', e.g. BALTOPS 2016 coverage (www.youtube.com/watch?v=vCugT4RmRpI), and 'NATO Maritime Power – Securing the Seas' (www.youtube.com/watch?v=0u27-sJ_OJ0).

References

68 *The Sea and International Relations*

References

Abulof, U. (2009). ' "Small peoples": the existential uncertainty of ethnonational communities', *International Studies Quarterly* 53(1): 227–48.

Alexander, J. C. (2010). 'Iconic consciousness: the material feeling of meaning', *Thesis Eleven* 103(1): 10–25.

Bartelson, J. (2014). *Sovereignty as Symbolic Form*, Abingdon: Routledge.

Bell, C. (2009 [1992]). *Ritual Theory, Ritual Practice*, Oxford: Oxford University Press.

Bell, D. S. A. (2003). 'Mythscapes: memory, mythology, and national identity', *British Journal of Sociology* 54(1): 63–81.

Blomkvist, N. (1998). *Culture Clash or Compromise? The Europeanisation of the Baltic Sea Area 1100–1400*, Visby: Acta Visbyensia.

Blomkvist, N. (2005). *The Discovery of the Baltic. The Reception of a Catholic World-System in the European North (AD 1075–1225)*, Leiden: Brill.

Bourdieu, P. (1989). 'Social space and symbolic power', *Sociological Theory* 7(1): 14–25.

Braudel, F. (1995). *The Mediterranean and the Mediterranean World in the Age of Philip II*, Berkeley, CA: University of California Press.

Carrión, J. (2013). *Bookshops*, trans. P. Bush, London: Maclehose Press.

de Carvalho, B., and H. Leira (2022). 'Challenging order at sea: the early practice of privateering', in B. de Carvalho and H. Leira, eds, *The Sea and International Relations*, Manchester: Manchester University Press.

Cassirer, E. (1946). *Language and Myth*, New York: Dover.

Dening, G. (1992). *Mr Bligh's Bad Language: Passion, Power and Theatre on the Bounty*, Cambridge: Cambridge University Press.

Dening, G. (2004). 'Deep times, deep spaces: civilizing the sea', in B. Klein and G. Mackenthun, eds, *Sea Changes: Historicizing the Ocean*, London: Routledge, 13–36.

Duke, S., and C. Gebhard (2017). 'The EU and NATO's dilemmas with Russia and the prospects for deconfliction', *European Security* 26(3): 379–97.

Eliade, M. (1959). *The Sacred and the Profane: The Nature of Religion*, New York: Harcourt, Brace and World.

Etzioni, A. (2000). 'Toward a theory of public ritual', *Sociological Theory* 18(1): 44–59.

Foucault, M. (1965), *Madness and Civilization: A History of Insanity in the Age of Reason*, New York: Pantheon Books.

Frühling, S., and G. Lasconjarias (2016). 'NATO, A2/AD and the Kaliningrad challenge', *Survival: Global Politics and Strategy* 58(2): 95–116.

Goody, J. (1961). 'Religion and ritual: the definitional problem', *The British Journal of Sociology* 12(2): 142–64.

Hodges, B., J. Bugajski and P. B. Doran (2018). *Securing the Suwałki corridor: Strategy, Statecraft, Deterrence, and Defense*, Washington, DC: Center for European Policy Analysis.

Joenniemi, P., ed. (1993). *Cooperation in the Baltic Sea Region*, London: Taylor & Francis.

Kant, E. (1935). *Bevölkerung und Lebensraum Estlands: Ein Anthropoökologischer Beitrag Zur Kunde Baltoskandias* [Estonia's Population and Habitat: An Anthropoecological Contribution to Knowledge of Baltic-Scandinavia], Tartu: Akadeemiline kooperatiiv.

Kiudsoo, M. (2016). 'Idatee valvurid' [The guardians of the *Austrvegr*], *Go Reisiajakiri* 63, https://reisikirjad.gotravel.ee/ajakiri/idatee-valvurid/ (accessed 10 February 2022).

Klein, B., and G. Mackenthun, eds (2004). *Sea Changes: Historicizing the Ocean*, New York and London: Routledge.

Kõiv, M. (2017). 'Pytheas, ookean ja Thule' [Pytheas, Ocean and Thule], *Vikerkaar* (October).

Kramer, F. D., and M. Nordenman (2016). 'A maritime framework for the Baltic Sea region', *Atlantic Council Issue Brief* (March): 1–11.

Lang, V. (2014). 'Eestlaste juured Eestimaal' [The roots of Estonians in Estonia], *Vikerkaar* (July).

Lange, H., B. Combes, T. Jermalavicius and T. Lawrence (2019). 'To the seas again: maritime defence and deterrence in the Baltic region', *ICDS Report* (April).

Lanoszka, A., and M. A. Hunzeker (2019). *Conventional Deterrence and Landpower in Northeastern Europe*, U.S. Army War College: Strategic Studies Institute, https://press.armywarcollege.edu/cgi/viewcontent.cgi?article=1380&context=monographs (accessed 10 February 2022).

Linklater, A. (2019). 'Symbols and world politics: towards a long-term perspective on historical trends and contemporary challenges', *European Journal of International Relations* 25(3): 931–54.

Loorits, O. (2000). *Meie, eestlased* [We, the Estonians], Tartu: Ilmamaa.

Mägi, M. (2018). *In Austrvegr: The Role of the Eastern Baltic in Viking Age Communication across the Baltic Sea*, Leiden and Boston, MA: Brill.

Mälksoo, M. (2021). 'A ritual approach to deterrence: I am, therefore I deter', *European Journal of International Relations* 27(1): 53–78.

Mead, G. H. (2015). *Mind, Self, and Society: The Definitive Edition*, Chicago: The University of Chicago Press.

Meri, L. (1999). Lecture of the President of the Republic at the University of Helsinki (Helsinki, 21 October).

Meri, L. (2000a). The President of the Republic of Estonia at the School of Slavonic and East European Studies (London, 8 March).

Meri, L. (2000b). 'The role of small nations in the European Union', President of the Republic in the University of Turku (Turku, 25 May).

Meri, L. (2008). *Hõbevalge: Reisikiri suurest paugust, tuulest ja muinasluulest* [Silver White: A Travelogue about the Big Bang, the Wind and the Ancient Poetry], Tallinn: Eesti Päevalehe AS.

Mikita, V. (2008). *Metsik lingvistika: Sosinaid kartulikummardajate külast* [Wild Linguistics: Whispers from the Village of Potato Worshippers], Tallin: Välgi metsad.

Mikita, V. (2013). *Lingvistiline mets: Tsibihärblase paradigma. Teadvuse kiirendi* [The Linguistic Forest: The Wagtail Paradigm. The Accelerator of Consciousness], Tallin: Välgi metsad.

Mikita, V. (2015). *Lindvistika ehk metsa see lingvistika* [Bird-istics or to the Woods with the Linguistics], Tallin: Välgi metsad.

Mikita, V. (2017). *Kukeseene kuulamise kunst: Läänemeresoome elutunnet otsimas* [The Art of Listening to the Chanterrelle: In Search for the Baltic-Finnic Sensibility], Tallin: Välgi metsad.

Murphy, M., F. Hoffman and G. Schaub, Jr (2016). 'Hybrid maritime warfare and the Baltic Sea region', University of Copenhagen Centre for Military Studies Report, Copenhagen: University of Copenhagen, 25 October.

NATO (2011). *Alliance Maritime Strategy*, www.nato.int/cps/ua/natohq/official_texts_75615.htm (accessed 14 June 2021).

NATO (2016). Warsaw Summit Communiqué, issued by the Heads of State and Government participating in the meeting of the North Atlantic Council (Warsaw, 8–9 July).

NATO (2018). Brussels Summit Declaration, issued by the Heads of State and Government participating in the meeting of the North Atlantic Council (Brussels, 11–12 July).

NATO (2019). 'NATO navies test readiness in Baltic Sea', 9 June, www.nato.int/cps/en/natohq/news_166717.htm (accessed 14 June 2021).

Oplatka, A. (2001). *Lennart Meri – Eestile elatud elu. Kahekõne presidendiga* [Lennart Meri – a Life Lived for Estonia. A Dialogue with the President], Tartu: Ilmamaa.

Ortner, S. B. (1973). 'On key symbols', *American Anthropologist* 75: 1338–46.

Rappaport, R. (1979). *Ecology, Meaning, and Religion*, Berkeley, CA: North Atlantic Books.

Raukas, A., and W. Stankowski (2010). 'The Kaali crater field and other geosites of Saaremaa Island (Estonia): the perspectives for a geopark', *Geologos* 16(1): 59–68.

Ringmar, E. (2012). 'Performing international systems: two East-Asian alternatives to the Westphalian order', *International Organization* 66(1): 1–25.

Spykman, N. J. (1944). *The Geography of the Peace*, New York: Harcourt, Brace and Co.

Stavridis, J. (2016). 'A new cold war deep under the sea', *The Huffington Post*, 28 October.

Steinberg, P. E. (1999a). 'Lines of division, lines of connection: stewardship in the world ocean', *Geographical Review* 89(2): 254–64.

Steinberg, P. E. (1999b). 'The maritime mystique: sustainable development, capital mobility, and nostalgia in the world ocean', *Environment and Planning D: Society and Space* 17(4): 403–26.

Steinberg, P. E. (1999c). 'Navigating to multiple horizons: toward a geography of ocean-space', *Professional Geographer* 51(3): 366–75.

Steinberg, P. E. (2001). *The Social Construction of the Ocean*, Cambridge: Cambridge University Press.

Sundstrom, I. (2017). 'An adequate NATO maritime posture: the missing element for deterring Russia', The Centre for International Maritime Security, 20 March.

Tynkkynen, N. (2013). 'The challenge of environmental governance in the network society: the case of the Baltic Sea', *Environmental Policy and Governance* 23(6): 395–406.

Tynkkynen, N. (2015). 'Baltic Sea environment, knowledge and the politics of scale', *Journal of Environmental Policy and Planning* 17(2): 201–16.

Valeri, V. (2014). *Rituals and Annals: Between Anthropology and History*, Chicago: Hau Books.

Van Bochove, C. (2007). 'Market integration and the North Sea system (1600–1800)', in H. Brand and L. Müller, eds, *The Dynamics and Economic Culture in the North Sea- and Baltic Region in the Late Middle Ages and Early Modern Period*, Hilversum: Uitgeverij Verloren, 155–69.

Weissman, C. (2015). 'Behind the strange and controversial ritual when you cross the equator at sea', *Atlas Obscura*, 23 October.

3

The white man and the sea? Gender, race and foundations of order

Halvard Leira and Benjamin de Carvalho

Understanding how the sea has been constructed and ordered throughout the centuries requires us not only to evoke and critique changing notions of space and time but demands a broader and deeper dive into underlying conceptions of order, such as race and gender.[1] As the introduction (de Carvalho and Leira, 2022, this volume) and the preceding chapters have shown, there is a strong need to interrogate taken-for-granted conceptions about the sea. And as ample literatures in the social sciences and humanities have demonstrated, gender and race are typically part and parcel of the naturalised ways in which we approach the world. The aim of this chapter is to explore if and how conceptions of gender and race have operated – and still operate – to normalise the relationship of domination of 'the white man' over the sea.

While we try to provide a relatively broad overview, we put particular emphasis on the last two centuries and the view from 'the West'. The spatial vantage point is partly a reflection of our own situatedness, but also of how 'the West' (or, more precisely, a selection of Northern Atlantic littoral states) superimposed itself on the world and tried to order the globe in its own image. We stress, though, that this should only be seen as a starting point, and we include alternative representations as examples of how the oceans can be constructed differently. The temporal delimitation is tied to the concepts we are exploring. Differentiation based on 'sex' is obviously an ancient phenomenon, but the language of 'sex' has changed dramatically over time and our current discourse on 'sex' and 'gender' is largely a product of the last two centuries (Towns, 2021). Differentiation based on categories of human collectives has had a different trajectory, but the whole language of 'race' and 'civilisation' is also largely a product of the last two centuries (Bowden, 2021; Yao and Delatolla, 2021).

At the linguistic level, the sea, the ocean and assorted synonyms to these terms display a bewildering variety of genus. There is, to our knowledge, no clear pattern in whether they are neuter, masculine or feminine. We return to the ambiguities this allows for below. Moving on to cultural constructions though, in the Western tradition the sea has tended to be sexed (and later gendered) feminine. The oceans have been seen as wild, unruly and seductive; qualities diametrically opposed to the orderliness and structure of idealised masculinity, and in need of subjugation to a male order. It should come as no surprise then that the deities of the oceans have often been men, capable of taming the forces of wind and water. When they are feminine, such as Medusa or the mermaids, they tend to embody the dangers or the seductiveness of the sea. The danger of the sea has also been underscored by its association with its unfathomable dark expanse or its sheer exoticness. From being the vast, mysterious and endless space surrounding known (European) lands, the oceans have become the space of all that which is juxtaposed against white masculine order.

The logical corollary of this representation of the sea is that the oceans in Western narratives have typically been seen as an arena for white masculine expression. The sea is a place where manly men go to do masculine things; the quasi-ritual space where boys went to become men. Fishing, rowing, sailing, fighting; these have all been cast as masculine activities, and jobs at sea have overwhelmingly been filled by men. There is, for instance, a higher percentage of female astronauts than of female seafarers.[2] The idea that women on board were an omen of bad luck, reproduced in numerous accounts of popular culture, underscored the sea as a place for men (Cordingly, 2007: 154). Likewise, leisure at sea involving physical activity, such as recreational fishing, sailing or wave surfing, has typically been associated with a masculine mastering of the elements.

Over the last two centuries, this masculine mastery has often been racialised in the Western imagination. The men conquering the sea have been presented as white, European (or, more specifically Anglo-American) men, and their historical domination of the oceans has been contrasted with an alleged oceanic passivity of other peoples from other places. This is as obvious in popular culture as in academic literature, with the sea-power vs. land-power tradition in geopolitics as the most explicit example from International Relations (IR). The point is only reinforced when we think of 'discoveries' beyond European shores. So-called discoveries are only labelled as such when recounting the experiences of white men. Earlier travels and contact across the seas are not labelled as such in the dominant Western discourse (see for instance Hofmeyr, 2007).

The sea can thus be seen as doubly gendered and racialised in this discourse. On the one hand, the sea itself is represented as a gendered and racialised space. On the other hand, the sea is a site of deeply gendered and racialised relations. These two levels obviously interact, as in the white man conquering the dark and exotic, feminine ocean. The sea as a site is equally obviously also shot through with class relations, often corresponding with rankings based on race and gender.

A superficial reading of our own introduction to this volume could see us as symptomatic of this broad, traditional Western understanding of the sea and what happens on and around it. In the introduction, we point out how the sea has played a role in global politics through what we referred to as processes of taming, controlling and traversing. While this language can obviously be read as gendered and racialised, our point was not to present a catalogue of possible ways to understand the sea, but to lay out some central tropes in how Western states (and IR literature) have approached the oceans to this date. This was meant not to limit our work to these dimensions, but as a starting point for future work on the sea. To the extent that the dominant naval powers of the last two centuries have been Western and IR has been a decidedly Western science, it should come as no surprise that these standard tropes have been gendered and racialised. If we were not looking beyond these standard tropes, however, important insights would be lost.

In this chapter, we thus try to destabilise some of these taken-for-granted representations of the sea, and in particular of the sea as a site of specific constructions of identity. We take as our starting point the assumption that any hegemonic identity-construct takes a lot of discursive work to uphold, and leaves cracks open for alternative constructs. While it is easy to pinpoint hegemonic white masculinity at sea, there must logically be alternative framings and narratives to explore. As Steven C. McKay has discussed and substantiated, elaborating on the work of R. W. Connell, we should at least expect to find complicit masculinity, subordinate masculinity and marginalised masculinity at sea. These masculinities are interwoven with relations of race and class (McKay, 2007; McKay and Lucero-Prisno III, 2012).

Challenging the traditional Western/Anglo-American representations, in expert discourse as well as in popular imaginations, is important. It is particularly important because the gendering and racialisation of the sea has led to forgetting and normalisation: forgetting of the many different ways in which gender and race has been intertwined with ocean life, and normalisation of an idealised Westernised and masculine approach to the sea. As we will demonstrate, discuss and reference below, whereas the forgetting has been gradually overcome, in particular through work in the fields of maritime history, maritime sociology and maritime anthropology, the extent and effects of naturalisation have only recently been subject to exploration. The study of gender, race and the sea has only recently started to move on from the phase of *locating* women and race (and different masculinities and racialisations), to the next steps of such research, exploring gender and genderings, race and racialisation more generally.

Considering how starkly gendered and racialised the sea has been in the Western imagination, and the work being undertaken in adjacent disciplines, it is remarkable how little work has gone into exploring gender, race and the sea in IR. This is all the more striking considering how one of the seminal early texts of feminist IR explicitly dealt with the seaside in its very title, *Bananas, Beaches and Bases*, and included incisive analysis of the intersection of gender and race, e.g. in cruise liner staffing and in and around military bases (Enloe, 1990).[3] However, from the vantage point of most of IR, the sea just *is*. Our task in this chapter is thus not to excavate explicitly gendered or racialised constructions of the sea in IR, but on the one hand to illustrate how the taken-for-granted existence of the sea in IR is built on broader cultural constructions, and on the other hand to explore some of the possibilities that follow from taking gender and race seriously in the analysis of the sea and international relations. We hasten to add that we are by no means providing a thorough feminist or post-colonial reading of the sea.[4]

What we offer is an analytical survey, primarily based on existing literature in neighbouring disciplines, with some primary source examples added. Based on this survey, we urge our fellow explorers of the ocean in IR to pay appropriate attention to questions of gender and race.

Our exploration comes in three parts. In the first part, we (1) briefly discuss the traditional, masculinised and Eurocentric view of the ocean as expressed in Western discourse over the last centuries. We then (2) discuss the literature locating women and race and deconstructing hegemonic white masculinity in and around the oceans, as it has developed over the last four decades. Finally, we (3) suggest avenues of research for IR, building on and pursuing further the insights from neighbouring disciplines.

The white man and the sea

In the Western discourse of the last two centuries, the sea, as noted, has been doubly gendered and racialised. Gendering and racialisation have been a feature of the sea as such, as well as of the interactions between humans and the sea. These two levels are obviously interrelated. Even so, it makes sense to separate them for analytical purposes. Before turning to specific representations of gender and race at sea, we thus start this section by discussing representations of the sea.

While there is no overall cross-linguistic pattern to whether the nouns describing oceans and the sea are gendered masculine, feminine or neutral, there is little doubt that the traits associated with these nouns have, in the West, typically been associated with femininity, disorder and the barbaric, as opposed to masculinity, the ordered and the civilised. The ocean has been seen as in fluid flux, disorderly, wild and seductive. Such associations with allegedly feminine traits are evident in the bible, as well as in Greek and Norse mythology (Helmreich, 2017: 31–4). By way of example, on his travels, Odysseus had to watch out both for the danger of the female sea monsters Scylla and Charybdis, and for the Sirens, whose alluring song led sailors to wreck their ships. The literature of the last two centuries likewise presents us with ample examples. The ocean has been described as strange and full of emotion, as in Herman Melville's *Moby Dick*: 'There is, one knows not what sweet mystery about the sea, whose gently awful stirrings seem to speak of some hidden soul beneath.' Likewise, Kate Chopin (in *The Awakening*) described a caring, almost caressing, ocean: 'The touch of the sea is sensuous, enfolding the body in its soft, close embrace.' In the recent fantasy-blockbuster *Eragon*, Chrisopher Paolini underscored the limitlessness of the sea: 'The sea is emotion incarnate. It loves, hates, and weeps. It defies all attempts to capture it with words and rejects all shackles. No matter what you say about it, there is always that which you can't.' And for all his claims that there was no symbolism in *The Old Man and the Sea*, it is hard not to see Hemingway presenting the ocean as a loved female: 'But the old man always thought of her as feminine and as something that gave or withheld great favors, and if she did wild or wicked things it was because she could not help them. The moon affects her as it does a woman, he thought.' As Susan F. Beegel (2002: 131) has argued, this is 'a story about the tragic love of mortal man for capricious goddess'.

The sea as a physical entity has been less obviously associated with race, and as elaborated in the introduction, this association has been more recent than the association with sex/gender. However, across the Western literary tradition, there is an abundance of references to the sea or the ocean as black or dark, often combined with other adjectives such as 'foreboding' or 'dangerous'. Nowhere was this analogy more obvious, perhaps, than in the white supremacist Lothrop Stoddard's (1921 [1920]) *The Rising Tide of Color Against White World-Supremacy*. The title sets the scene, and nautical references abound in the text. Stoddard contrasted how European medieval 'forefathers' saw the ocean as 'a numbing, constricting presence; the abode of darkness and horror' (1921 [1920]: 147), but how Columbus changed all this and allowed the Europeans to dominate by 'rising from the pathless ocean' (1921 [1920]: 148). At the end of the First World War, the 'white world' nevertheless faced the need to take drastic measures. 'Above all: time presses, and drift is fatal. The tide ebbs. The swimmer must put forth strong strokes to reach the shore. Else – swift oblivion in the dark ocean' (1921 [1920]: 196–7). What had to be avoided as a solution, though, was Bolshevism. If that ideology was successful, the white race 'would sink like lead into the depths of degenerate barbarism' (1921 [1920]: 219). The juxtaposition is similar to the gendered one – what confronts the civilised white man is an ocean of dark barbarism.

As Helmreich (2011: 134) discusses, the representation of the oceans differs significantly from this in much early anthropology. There, seawater was much lighter: blue, green, grey or gleaming. Following Helmreich, the oceans nevertheless functioned solely as nature, juxtaposed against the theorisation which could start only on land. As summarised by DeLoughrey (cited in Helmreich, 2011: 134–35), 'colonial mystifications of an idyllic South Seas … interpellated the Pacific Basin as a vast, empty (feminized) ocean to be filled by masculine European voyagers'.

Underlying constructions of the sea as gendered feminine and coded black (or at least exotic) is the basic juxtaposition of sea and land, mirroring the contrast between nature and culture. As demonstrated in feminist and post-colonial scholarship over decades, the former is typically gendered feminine and/or barbarian (that is, non-white) and the latter is typically gendered masculine, civilised and white (see MacCormack and Strathern, 1980; Helmreich, 2017). Land can be cultivated and ordered, while the sea can at best be mastered for a fleeting moment.[5] Turning to basic metaphors, land has, at least since the origin of early modern map-making in Europe, been perceived as a rationally bounded space (Branch, 2014), while the sea has been seen as irrational and boundless; 'hic sunt dragones' indeed. On yet another basic metaphor, land ties in with 'up', as in heights, whereas the ocean ties in with 'down', as in depth. In the Western tradition, the former is clearly privileged over the latter. All of these dichotomies work together to establish the ocean as feminine and coloured and subordinate to landed space and white men. Within the discipline of IR, the sea and the ocean has typically not been ascribed any particular quality. As we argue in the introduction to this volume, it has been seen as something to traverse or exploit. Even so, and perhaps even more strongly in IR than in other social sciences, the gendered division between sea and land, nature and culture has structured how the sea and interaction with it has been perceived, as in the above-cited tradition of geopolitics and later in strategic studies and security studies.

As several of the examples above illustrate, the overall tendency to gender the sea as feminine has been intertwined with a masculine gendering of the handling of the sea. Although the sea has been gendered as feminine, femininity in the Western socio-economic order has been strongly associated with landed activities. Chief oceanic deities like Poseidon, Neptune and Njord were men, and in the bible, Yahve has the power to create the flood and to make it recede. The ocean has been alluring to men, who have 'One foot in sea, and one on shore' (Shakespeare, *Much Ado about Nothing*), and the way to deal with it has been to tame it, joust with it, to metaphorically 'take arms against a sea of troubles' (Shakespeare, *Hamlet*) or to die trying. Here there is great continuity across centuries, in an ethnography of surfers, Waitt (2008: 92) summarises the masculine surfer view of the sea:

> The ocean, scripted as an enemy, is respected, but is also slashed, cut, smashed and eventually killed by skilled manoeuvres. In this account, masculinity is produced through the bodily practices of daily surfing routines, as prized breaks are to be 'defended' from rogue surfers or 'tamed' through surviving near-death experiences of drowning.

Surfing aside, the technological advances of steam shipping (and later nuclear propulsion) and aeroplanes has allowed for a general assumption that the ocean has been mastered, allowing for the flattening of ocean space as described in the introduction to this volume. Thus, it has until recently been possible for IR to conceive of oceans as something to secure domination, control or command over (e.g. Modelski and Thompson, 1998), or as something to be rationally governed through cooperative regimes (e.g. Stokke, 2007).

The masculine handling of the sea has also typically been coloured white, often retroactively, over the last two centuries. In early twentieth-century Europe, the story of European expansion around the globe was told as a story of the success of the white man in traversing the oceans no one else could cross, and in applying superior seafaring skills, ingenuity, bravery and fighting prowess in overcoming all others who took to the sea. This initial mastery of the sea allegedly had allowed for the further deployment of Western domination over other people and the constitution of worldwide European empires.[6] As UK historian and navy officer G. A. Ballard put it for an Indian Ocean naval context in 1927 (Ballard, 1998 [1927]: 59), 'Vasco da Gama showed that on the scientific side of war the white man was far in advance of the brown ... on the moral side of war the superiority of the white man was even more pronounced.' This held true to such an extent that even in advantageous situations, the 'Asiatic ... behaved after the fashion of the wild creature which encounters a human being in the jungle, whom it could kill with ease, but flies from under the influence of an uncontrollable awe' (Ballard, 1998 [1927]: 60). Whereas the tone has changed, much later scholarship has followed the same tropes of European domination at sea and non-European subordination. The North Africa based so-called 'Barbary pirates' of the seventeenth to nineteenth centuries can serve as an example. There had been intense fighting between Christian and Muslim forces in the western Mediterranean since the early sixteenth century. However, it was only when Dutch and English privateers turned renegade and introduced 'round ships' with square rigging that the alleged 'pirates' became a menace in the Atlantic (see e.g. Wadsworth, 2019: 141). As the story goes, it took the help of unfaithful Europeans for non-Europeans to threaten European mastery of the sea.

A corollary to the white masculine mastering of the sea has been the construction of the seafarer as somewhat of a liminal figure, a man living on the edge between civilisation and chaos. Much like the cowboy, he would spend most of his time away from the leisure of landed life, returning to the (liminal) port at irregular intervals to continue his conquests. Thus, stories of seafarers, both prose and fiction, historical and current, have often included reference to sexual exploits and prostitutes.

Locating intersectionalities: gender/race, sea/land

As we noted in the introduction to this chapter, the established constructs of the sea and life on it have been challenged repeatedly in academic scholarship over the last decade. The first step of many feminist academic endeavours has been to locate the women, to show how naturalised masculinity glosses over female presence and experience (Foot, 1990; Towns, 2021). In parallel, scholars explore how seemingly hegemonic masculinities always open up room for other forms of masculinity (Connell and Messerschmidt, 2005). Much the same has been the case for post-colonial studies – a first step has been to demonstrate that the whiteness of history is the result of an unconscious (or sometimes conscious) disregard for the agency of non-white actors (Hall, 1981; Yao and Delatolla, 2021). Studies of the sea are no different, and as we elaborate in this section, the efforts to write women, alternative masculinities and race into maritime history and other studies of the oceans stretch back several decades.

Let us return to the dominant representations of the sea. While the dominant constructions have been of the sea as dark, mysterious, dangerous and feminine, there is obviously nothing inherent in these constructions. In Inuit Thule culture, archaeologists and ethnographers have for instance described a very clear dichotomous distinction between land and sea, coded as male and female, and associated with summer vs. winter and antler vs. ivory. However, there seems to have been no necessary ranking of the two (Hodder and Hutson, 2003: 58). There is also an ambiguity inherent in the oceanic pantheon. Even if most ocean deities have been male, in the Catholic faith, Virgin Mary is seen as the protector of sailors and fishermen.

Furthermore, as alluded to above, even the basic constructions of the sea and seafarers are potentially polyphone and dominant depictions also have their alternatives. The oceans are both alluring and terrifying and have at times been associated with masculine traits as well as feminine ones. If we return to the example of *Moby Dick*, it should be noted that Melville did not see the ocean as uniformly feminine. He described the 'strong, troubled, murderous thinkings of the masculine sea', and the explicit gendered representations of the ocean in *Moby Dick* tend to be masculine rather than feminine (see Beegel, 2002). As Rita Bode (2006) has argued, there is an ambiguity to this gendering in *Moby Dick*, expressed in terms of masculinity and maternity, and it could be argued that the masculinity is more about how to handle the sea than about the ocean itself. Both in Melville and the surfing quote above, the ocean can be understood as a respected or worthy opponent, which would typically imply masculinity.

From the character of the sea, we now move to life on the sea, starting with the very basic question: 'where are the women?' At a general level, much academic effort has gone into challenging the naturalised gendered view of the sea and in particular life on and around the sea (see overviews in Davis and Nadel-Klein, 1992; Stanley, 2003; Colville et al., 2015; Gissi et al., 2018). Klein and Mackenthun (2004: 4) put it well: 'To see the ocean and the social space of the ship as exclusively male spheres of action is to ignore both the many women who traveled on board ship in various capacities, and the ways in which the life of seafaring men interacted with that of women on shore.' As demonstrated in numerous fields, women have not been absent, they have simply been excluded from the naturalised narratives; 'even when women were there at sea they were not recorded as there' (Stanley, 2003: 135). While the early efforts to rectify this were focused on individual women (such as the female pirates Anne Bonney and Mary Read), the next steps in locating women involved studying broader industries, such as fisheries (see Kaplan, 1988 for an early example) and the passenger liner/cruise industry (Coons, 2008). Gradually the focus has moved to exploring the intersection of gender and the sea more generally. Although important individual stories have been told about women at sea, it remains true that there has been an enormous majority of men at sea. As McKay (2007: 619) summarises the scholarship on the topic, 'Few occupations can match seafaring for its distinct and historically gendered identity, which seems to date from the very inception of merchant shipping in the Mediterranean and in East and South-East Asia.' Gender studies of the maritime world have thus widened the scope, drawing in ports, the overall naval economy and varieties of masculinity at sea.

Thus, even in literatures locating women, there has been a tendency to look beyond the sea in a narrow sense. In fisheries for instance, women have been shown to be integral to the overall fish economies, not only through the entire supply chain of fish, but also by taking up traditional 'masculine' tasks on land, thus making it possible for men to leave for months on end. As poignantly stated, *Captain Ahab Had a Wife* (Norling, 2000). Likewise, the social and gender historian Ragnhild Aarsæther (1981) commented that the Norwegian fisherman-farmer, so called because neither farming nor fishing could be year-round activities, was a fisherman at sea and a wife who worked the land. Closer studies of seafaring communities in many different locations have demonstrated similar patterns – women who have had their men at sea for prolonged periods have been central cogs in family, society and the economy (see e.g. Hagmark-Cooper, 2016; Tenold and Økland, 2019). Furthermore, as soon as one looks outside the modern economies of the West, women show up as fisherfolk themselves (Davis and Nadel-Klein, 1992: 143–45). This latter observation drives home what should be an obvious point, namely that the gendering of land-based and sea-based labour is socially constructed and that such labour is not necessarily tied closely to notions of gender at all.

Furthermore, as in many other instances, looking beyond the Western experience we find alternative genderings both of the sea and of seafarers. Anthropologists have for instance discussed numerous examples of how fishing has been an activity for both genders, or a female speciality (McDowell, 1984; Errington and Gewertz, 1987; see also the more general overview in Frangoudes and Gerrard, 2019). Matrilineal organisation of work and female oceanic deities can also be juxtaposed against the Western construction of the sea, although the alternative genderings studied have not implied a more general reversal of gender roles.

Unsurprisingly, given the liminality of the mariner, a lot of attention has been paid to the ones going beyond the pale, to piracy.[7] This has included explorations of gender. But while there are historical examples of female pirates (Duncombe, 2017), these tended to be exceptions rather than the rule (Appleby, 2013). Women were, however, clearly involved in the broader piratical economies, as victims, partners, in prostitution and as sponsors. Gender-wise, significant attention has also been paid to another question than the possible inclusion of women, namely alternative genderings. Piracy involved radically different forms of life and organisation than what could be found on land or on other ships, and it has been speculated that this might have opened up space for different ways of thinking and practising masculinity. B. R. Burg (1983) infamously argued that sodomy was a common feature among pirates. The scarcity of source material has made this hard to substantiate, and later queer pirate studies have focused more on representations than on actual queerness (see the discussion in Marsellas, 2018). Hans Turley (1999) for instance suggested that the masculinities of pirates were presented as ambiguous in the contemporaneous literature, and that this has been part of what has made pirates both feared and romanticised.[8]

Regardless of the progress made in locating women on and around the historical seas, locating women on the contemporary seas is still a challenge. As mentioned in the introduction, the International Maritime Organization (IMO) estimates that only 2 per cent of seafarers are female. That number, easily available on the IMO website, restated on the International Transport Worker's Federation website and repeated in news stories at intermittent intervals as well as in current research (MacNeil and Ghosh, 2017), is nevertheless relatively poorly substantiated. It dates to a 1992 report by the IMO (Belcher et al., 2003: 9) and seems to have taken on a life of its own since then, without any additional or more recent data cited to support it. It is hard to overcome both gender imbalance and the stereotype of gender imbalance, if the data are three decades old. Some newer data have recently been brought forward, suggesting that the percentage is closer to 1 (although importantly, those data exclude the cruise industry – as does the IMO report). In short, no one knows for sure the female percentage of world seafarers, but if the cruise industry numbers are factored in, a guesstimate of 5 per cent seems more likely than the often-reported 2 per cent.[9] The point here is not to downplay the extremely obvious gender gap in the maritime industries, but to highlight how the traditional counting statistics continue to downplay or even ignore the fields where women for decades have been working at sea.

Similarly, the ease with which gender stereotypes are naturalised in reference to the maritime world might undermine the very project of equality. To take but one example, MacNeil and Ghosh write with the obvious intention of reducing the gender gap in the maritime industries. However, when describing the roots of this gender gap, they for instance argue that 'The superstition that it is bad luck to have women on board ships greatly decreased the seafaring opportunities for women over the centuries' (MacNeil and Ghosh, 2017: 43). While there is no denying the existence of this myth, and it seems likely that it has justified instances of gender exclusion, considering the number of women crossing the seas as migrants, slaves or wives (see Cordingly, 2007: 154–55), it is untenable to uphold it as a reason for the maritime gender gap. If women were at sea in droves, and they were, albeit more often than not in passive roles, the superstition that they brought bad luck can't explain why they have not been part of the maritime workforce. Industry-specific prejudices and path-dependencies over the last century would seem to hold greater explanatory power than centuries-old myths with an unclear relationship to maritime practice. The construction of women as more family-oriented and home-bound has for instance been used as a reason for excluding women from the labour market in general, and it has had added salience in maritime employment, given the long stretches spent at sea. However, as demonstrated in recent research, there is no reason to expect men and women to react differently to the stress of separation, and prolonged separation is a common reason for men leaving a maritime career (Thomas, 2004: 315–16).

When questioning the alleged 'whiteness' of life at sea, maritime history and maritime sociology and anthropology have carried out a two-pronged attack, highlighting both the naval activities of non-Western people and the many interrelations between different cultures. To return one last time to *Moby Dick*, Melville also presents us with ambiguities of race, for instance in the multiethnic crew of the *Pequod* and in the whiteness of the terrifying whale (see Bernard, 2002). In a much more recent exploration, de Almeida (2004: x) stressed how the Southern Atlantic which he observed did not confirm to Western expectations of tropical bliss; rather it was 'unpleasantly brown, muddy, earth-colored'. To him, this became a metaphor for the opaqueness of the sea and its many sedimented histories. Unsurprisingly, history turns out to have been many-coloured. Going back before recorded time, the spread of mankind to the islands of the Pacific easily put lie to the later Western thesis of an exclusive European (i.e. 'white') mastery of the sea. Moving into the early modern period, it has been demonstrated thoroughly how the Ming empire in China engaged in massive naval expeditions in the Indian Ocean, merely decades before the Europeans arrived on much smaller vessels at a much slower rate. As to the point raised above about the 'Barbary pirates', while the square rigging was imported from Europe, the fighting skills needed to win hand-to-hand combat was still provided by the Ottoman Janissary contingents on board (themselves a product of the polyglot Ottoman Empire). These were truly joint ventures, combining land and sea and different cultures. Increasingly it has come to be accepted that a striking feature of Mediterranean/ North-African interactions from the seventeenth to the nineteenth centuries was the degree of creativity and accommodation which went into intercultural relations (Windler, 2001). Likewise, crew rolls of captured pirate ships have long demonstrated the varied background of the pirates, and when slave ships were captured by pirates, it was not uncommon for the slaves to join the pirate crew as equals (Kinkor, 2001). As Linebaugh and Rediker (2000) demonstrated more broadly, multiethnic and 'multiracial' crews were common in the seventeenth and eighteenth-century Atlantic.

Several different histories have thus stressed how seafaring was not a particularly European speciality. A different take on race and the oceans has been provided with growing recognition over recent decades of the importance of the Atlantic slave trade to the growth of European empires and the European expansion more generally. Considering that slaves usually outnumbered crew by a factor around ten, it seems at least possible that between the late seventeenth century and the middle nineteenth century, a majority of the people traversing the high seas were non-white. In the whitewashing of the oceans, the Middle Passage was for a long time completely ignored.[10] The effects of slavery and the slave trade have gradually been incorporated in the study of the oceans, leading, for example, to the development of the idea of the *Black Atlantic* (Gilroy, 1993), a black identity not tied specifically to one territory but built on shared experiences across the sea (and of having crossed the sea in the hull of slave ships).

Considering how men outnumber women at sea, and the multinational background of seafarers, an exploration of multiple masculinities at sea offers a particularly fruitful avenue for questions about gender and race at sea. While the handling of the oceans, as described above, has had an overall white male framing, allowing for an exemplary masculinity, life on the ocean has been a lot less clear-cut. The strict hierarchies on most ships have implied that the masculinities of lower-ranked seafarers at sea have been both marginal and subordinate. On the other hand, when in port or at home, seafarers have often been construed (and self-identified) as examples of hyper-masculinity or patriarchal values. As Steven McKay (McKay, 2007; McKay and Lucero-Prisno III, 2012) has elaborated, using the Filipino seafarers as his example, multiple masculinities are in play and must be navigated.[11] Ethnicity, gender and class interact, both on ship, in port and at home. To the Filipino state, the seafarers are heroic in their work for remittances; to the European officers of Filipino ratings they can be construed as effeminate or homosexual; among themselves on board they ascribe to a marginal, yet prideful and creative masculinity; in port they might combine gallantry and sexual exploits; while at home they are construed as providers and patrons. Adding further complexity, with the growing number of Filipino women working in the cruise industry, the notion of the heroic seafarer has become a lot less gender-specific over the last decades. In making sense of the sea and gender, it is the intersection of these *multiple genderings* with race that hold the most promise for future scholarship.

Possible futures

Gender and race have mattered and continue to matter in how the oceans are constructed and how the relations between humans and the oceans and between humans on the oceans play out. As discussed above, the ways in which they interact and intersect clearly show the multiple orderings upon which the dominant discourses of the sea rely. Gender and race are therefore key to the international relations of the sea. Taming, traversing and controlling are not simply a list of ways states have approached the sea; they are a catalogue of ways in which the sea and relations between state and the sea have been constructed in gendered and racialised ways. Having established that the construction of the oceans as the domain of the white, heterosexual man is contingent, we now move on to some possible implications of taking gender and race seriously in analysing international relations and the sea.

As a starting point, the mere act of thinking seriously about gender and race opens new perspectives on international relations and the sea. At the macro level, we noted above how an important strand of geopolitics was founded on a partially racialised understanding of the juxtaposition between land and sea. At the micro level, the gendered division of labour in the cruise industry was discussed by Enloe more than three decades ago. The construction of the sea as a space to be conquered or mastered has obviously tied in well with a focus on maritime security in IR more generally, to the extent that many other aspects of oceanic life, such as the ones Enloe discussed, have been ignored. Thinking about the oceans in less masculine ways might lead to changes in academic priorities.

In economic experiments in local fisheries, women have been shown to produce more sustainable catches than men, and to be quicker to change to lower resource extraction (Gissi et al., 2018: 216–17). Even so, one should obviously avoid the essentialised notion that women are predisposed or better suited to saving the globe (Gissi et al., 2018: 216–17). Rather than seeing a gender difference in approaches to the ocean, findings such as these should lead us to question the taken-for-granted constructions of land and sea.

Beyond the gendering of oceans, there is no denying the massive gender discrepancy at sea, historically as well as contemporarily. However, as demonstrated repeatedly in adjacent disciplines, the overall oceanic economies, from piracy to fisheries, have relied on a combination of sea and land, and on female contributions on land. This point dovetails with the feminist critique of security studies. In a sense, it is a restatement of Enloe's point, but it must be repeated – the heavily masculinised activities at sea have relied and continue to rely on landed labour, where women have been just as important as men. Including gender perspectives should in this perspective help to denaturalise some of the divisions between oceanic and landed activities.

Taking race into account offers some of the same opportunities for rethinking and should lead to a reevaluation of the historical international relations of the oceans. This is already under way. Increasing attention has been paid to the creative cooperation between Europeans and a variety of local actors in Africa and Asia in the early centuries of European expansion (Phillips and Sharman, 2020). This suggests a moving away from the focus on European mastery of the waves, incorporating a view of military force as amphibious and a highlighting of the interrelations between sea and land, European and local.

Similarly, there has been some interest in slavery and the slave trade within an IR framework. Almost two decades ago, Barkawi and Laffey (2002: 113–14) suggested incorporating insights from Gilroy's notion of the Black Atlantic, particularly how the modern world was built in part on the slave trade, in the analyses of international relations past and present. This would also necessitate an engagement with empires and imperialism. Since then, there has been an impressive growth in post-colonial and decolonial studies in and around IR. However, explicit engagements with the ocean in general (with the notable exception of Shilliam's (2015) exploration of *The Black Pacific*; see also de Almeida, 2004; Tinsley, 2008) and slavery and the slave trade more specifically have been lacking. To the extent that these topics have been explored in IR, it has primarily been in discussions of the campaign to abolish the slave trade, with abolitionism used as a case study (Ray, 1989; Kaufmann and Pape, 1999). An example closer to our concerns here, incorporating the relationship between sea and land and Europeans and local actors, can be found in Keene's (2007) study of the British treaty-system against the slave trade. The bigger story of the international relations of (the transatlantic) slave trade and slavery nevertheless remains unwritten.

In current affairs, the questions of race and nationality have been largely ignored in IR. Our examples concerning Filipino mariners suggest one reason why this is untenable. There is a glaring discrepancy between the nations with the biggest navies and merchant navies and the countries supplying most seafarers. This would seem to be a topic ripe for study by IR scholars concerned with the intersection of politics, economy and identity, including scholars concerned with traditional and non-traditional security challenges, diaspora politics and global development. While it is perfectly possible to study the sea and international relations (or the sea through the lenses of IR) without engaging with gender and/or race, we hope that the preceding pages will have convinced that analyses including these topics can only help to broaden our understanding of both the sea and relations on and around it and help fully make sense of the multiplicity of ways in which the sea impacts international relations.

Notes

1 We would like to thank the anonymous reviewers for MUP for the impetus to write a chapter tackling these topics head on, and Ann Towns for her very helpful comments on an earlier draft. All errors and omissions remain our responsibility.

2 The International Maritime Organization estimates that, despite three decades of dedicated gender programmes, 'women represent only two percent of the world's 1.2 million seafarers and 94 percent of female seafarers are working in the cruise industry' (IMO, no date). Around

11 per cent of the people who have been in space ('astronauts' for the
sake of brevity) have been women (Big Think, 2018).

3 It should be added, though, that the title's focus on beaches must be
understood as primarily alliterative. While Enloe discussed tourism
in detail, beaches and beach activities were not central parts of that
discussion.

4 Such thorough readings would necessitate more space than what the
current format allows, and a more systematic engagement with primary sources.

5 This should not be read as an assertion that the relationship between
the sea and order is monolithic. Rather, as Steinberg (2001: 207–8)
demonstrated, ocean space has been constructed in different ways as
a) non-territorial/asocial, b) territorial and an extension of land space
and c) a non-possessable 'force-field'. The first and third of these
constructions, which are by far the most common, map well onto a
masculine/feminine dichotomy.

6 It bears repeating that the importance of skin colour as a differentiating
feature changed greatly between the early travels of exploration in the
late fifteenth century and the heyday of European imperialism in the
nineteenth and twentieth centuries, and that the discourse about race is
a phenomenon of the last two centuries.

7 The interest in historical piracy, which goes back some four decades
(Ritchie, 1986; Rediker, 1987), got an added impetus from the piracy
outside of Somalia around 2010 (Policante, 2015).

8 It is also important to remember that the constructions of gender and
gender relations common to the early twenty-first century do not cor-
respond to constructions of gender before the nineteenth century. And
for obvious reasons we really do not know much about the prevalence
of homosexual practice in homosocial environments. Downs (2020)
provides an illuminating account from an 1840s British prison ship in
Bermuda, but we have no way of knowing how representative it was.

9 BIMCO/ICS put the number of seafarers (excluding the cruise industry)
in 2015 at 1.65 million, and through a sample suggested that 1 per
cent of those were women (thus around 16,500) (ILO, 2019: 32). The
cruise industry claims to employ around a quarter of a million seafarers
(Marcellin, 2019), of which ITF estimates that just short of one third are
women (thus around 80,000) (ILO, 2019: 32). The overall number of
women seafarers could thus be estimated at just short of 100,000, out
of a total number of some 1.9 million, equalling 5 per cent.

10 We have mentioned above the common myth that it meant bad luck
to have women aboard, and how it has probably been less about femininity than about prejudice against women actively partaking in the

The white man and the sea?

maritime workforce. The alleged myth was obviously no hindrance to
having female slaves in the hold.

11 The Philippines is estimated to be the second largest seafarer supply
country in the world, the largest for ratings.

References

Aarsæther, R. (1981). 'Kvinner som familieforsørgere og sjølforsørgende
belyst ved nordnorske skattelister på 1500–1600-tallet', in H. Gunneng
and B. Strand, eds, *Kvinnans ekonomiska ställning under nordisk
medeltid*, Göteborg: Lindome, 41–9.

de Almeida, M. V. (2004). *An Earth-colored Sea: 'Race', Culture, and the
Politics of Identity in the Post-Colonial Portuguese-Speaking World*,
New York: Berghahn Books.

Appleby, J. C. (2013). *Women and English Piracy 1540–1720: Partners
and Victims of Crime*, Woodbridge: Boydell Press.

Ballard, G. A. (1998 [1927]). *Rulers of the Indian Ocean*, New Delhi: AES
Reprint.

Barkawi, T., and M. Laffey (2002). 'Retrieving the imperial: empire and
International Relations', *Millennium* 31(1): 109–27.

Beegel, S. F. (2002). 'Santiago and the eternal feminine: gendering *La
Mar* in *The Old Man and the Sea*', in L. R. Broer and G. Holland,
eds, *Hemingway and Women: Female Critics and the Female Voice*,
Tuscaloosa, AL: University of Alabama Press, 131–55.

Belcher, P., H. Sampson, M. Thomas, J. Viega and M. Zhao (2003).
Women Seafarers. Global Employment Policies and Practices, Geneva:
International Labour Office.

Bernard, F. V. (2002). 'The question of race in *Moby-Dick*', *The
Massachusetts Review* 43(3): 384–404.

Big Think (2018). https://bigthink.com/think-tank/women-in-space-by-
country (accessed 7 December 2020).

Bode, R. (2006). ' "Suckled by the sea": the maternal in *Moby-Dick*', in
E. Schultz and H. Springer, eds, *Melville and Women*, Kent, OH: Kent
State University Press, 181–98.

Bowden, B. (2021). 'Eurocentrism and civilization', in B. de Carvalho,
J. Costa López and H. Leira, eds, *The Routledge Handbook of Historical
International Relations*, Abingdon: Routledge, 162–70.

Branch, J. (2014). *The Cartographic State: Maps, Territory, and the Origins
of Sovereignty*, Cambridge: Cambridge University Press.

Burg, B. R. (1983). *Sodomy and the Pirate Tradition. English Sea Rovers in
the Seventeenth Century Caribbean*, New York: NYU Press.

de Carvalho, B., and H. Leira (2022). 'Introduction: staring at the sea', in B. de Carvalho and H. Leira, eds, *The Sea and International Relations*, Manchester: Manchester University Press.

Colville, Q., E. Jones and K. Parker (2015). 'Gendering the maritime world', *Journal for Maritime Research* 17(2): 97–101.

Connell, R. W., and J. W. Messerschmidt (2005). 'Hegemonic masculinity. rethinking the concept', *Gender and Society* 19(6): 829–59.

Coons, L. (2008). 'From "company widow" to "New Woman": female seafarers aboard the "floating palaces" of the interwar years', *International Journal of Maritime History* 20(2): 143–74.

Cordingly, D. (2007). *Seafaring Women. Adventures of Pirate Queens, Female Stowaways, and Sailor's Wives*, New York: Random House.

Davis, D. L., and J. Nadel-Klein (1992). 'Gender, culture, and the sea: contemporary theoretical approaches', *Society and Natural Resources* 5(2): 135–47.

Downs, J. (2020). 'The gay marriages of a nineteenth-century prison ship', *The New Yorker*, 2 July, www.newyorker.com/culture/culture-desk/the-gay-marriages-of-a-nineteenth-century-prison-ship (accessed 31 August 2020).

Duncombe, L. S. (2017). *Pirate Women. The Princesses, Prostitutes, and Privateers who Ruled the Seven Seas*, Chicago: Chicago Review Press.

Enloe, C. (1990). *Bananas, Beaches, and Bases: Making Feminist Sense of International Politics*, Berkeley, CA: University of California Press.

Errington, F. K., and D. Gewertz (1987). *Cultural Alternatives and a Feminist Anthropology: An Analysis of Culturally Constructed Gender Interests in Papua New Guinea*, Cambridge: Cambridge University Press.

Foot, R. (1990). 'Where are the women? The gender dimension in the study of international relations', *Diplomatic History* 14(4): 615–22.

Frangoudes, K., and S. Gerrard (2019). 'Gender perspective in fisheries: examples from the South and the North', in R. Chuenpagdee and S. Jentoft, eds, *Transdisciplinarity for Small-Scale Fisheries Governance. Analysis and Practice*, Cham: Springer, 119–40.

Gilroy, P. (1993). *The Black Atlantic: Modernity and Double Consciousness*, London: Verso.

Gissi, E., M. E. Portman and A.-K. Hornridge (2018). 'Un-gendering the ocean: Why women matter in ocean governance for sustainability', *Marine Policy* 94: 215–19.

Hagmark-Cooper, H. (2016). 'To be a sailor's wife: ideals and images of the twentieth-century seafarer's wife in the Åland Islands', in B. Beven, K. Bell and R. James, eds, *Port Towns and Urban Cultures: International Histories of the Waterfront, c.1700–2000*, London: Palgrave Macmillan, 221–42.

Hall, S. (1981). 'The whites of their eyes: racist ideologies and the media', in G. Bridges and R. Brunt, eds, *Silver Linings: Some Strategies for the Eighties*, London: Lawrence and Wishart, 28–52.

Helmreich, S. (2011). 'Nature/culture/seawater', *American Anthropologist*, 113(1): 132–44.

Helmreich, S. (2017). 'The genders of waves', *Women's Studies Quarterly* 45(1–2): 29–51.

Hofmeyr, I. (2007). 'The black Atlantic meets the Indian Ocean: forging new paradigms of transnationalism for the Global South – literary and cultural perspectives', *Social Dynamics* 33(2): 3–32.

Hodder, I., and S. Hutson (2003). *Reading the Past: Current Approaches to Interpretation in Archaeology*, 3rd edn, Cambridge: Cambridge University Press.

International Labour Organization (ILO) (2019). *Recruitment and Retention of Seafarers and the Promotion of Opportunities for Women Seafarers*, Geneva: ILO.

International Maritime Organization (IMO) (no date). 'Women in maritime', www.imo.org/en/OurWork/TechnicalCooperation/Pages/WomenIn Maritime.aspx (accessed 7 December 2020).

Kaplan, I. M. (1988). 'Women who go to sea: working in the commercial fishing industry', *Journal of Contemporary Ethnography* 16(4): 491–514.

Kaufmann, C. D., and R. A. Pape (1999). 'Explaining costly international moral action: Britain's sixty-year campaign against the Atlantic slave trade', *International Organization* 53(4): 631–68.

Keene, E. (2007). 'A case study of the construction of international hierarchy: British treaty-making against the slave trade in the early nineteenth century', *International Organization* 61(2): 311–39.

Kinkor, K. J. (2011). 'Black men under the Black flag', in C. R. Pennell, ed., *Bandits at Sea: A Pirates Reader*, New York: NYU Press, 195–210.

Klein, B., and G. Mackenthun (2004). 'Introduction: the sea is history', in B. Klein and G. Mackenthun, eds, *Sea Changes: Historicizing the Ocean*, London: Routledge, 1–12.

Linebaugh, P., and M. Rediker (2000). *The Many-headed Hydra: Sailors, Slaves, Commoners, and the Hidden History of the Revolutionary Atlantic*, London: Verso.

MacCormack, C. P., and M. Strathern, eds (1980). *Nature, Culture and Gender*, Cambridge: Cambridge University Press.

MacNeil, A., and S. Ghosh (2017). 'Gender imbalance in the maritime industry: impediments, initiatives and recommendations', *Australian Journal of Maritime and Ocean Affairs* 9(1): 42–55.

Marcellin, F. (2019). 'How growth in the cruise industry is causing a drought in recruitment', www.ship-technology.com/features/cruise-industry-recruitment-challenges/ (accessed 27 August 2020).

Marsellas, N. (2018). 'Swashbuckling sexuality: the problem with queer pirates', in A. Sanna, ed., *Pirates in History and Popular Culture*, New York: McFarland, 50–60.

McDowell, N. (1984). 'Complementarity: the relationship between female and male in the East Sepik village of Bun, Papua New Guinea', in D. O'Brien and S. Tiffany, eds, *Rethinking Women's Roles: Perspectives from the Pacific*, Berkeley, CA: University of California Press, 32–52.

McKay, S. C. (2007). 'Filipino sea men: constructing masculinities in an ethnic labour niche', *Journal of Ethnic and Migration Studies* 33(4): 617–33.

McKay, S. C., and D. E. Lucero-Prisno III (2012). 'Masculinities afloat: Filipino seafarers and the situational performance of manhood', in M. Ford and L. Lyons, eds, *Men and Masculinities in Southeast Asia*, New York: Routledge, 20–37.

Modelski, G., and W. R. Thompson (1998). *Seapower in Global Politics, 1494–1993*, Houndmills: Macmillan.

Norling, L. (2000). *Captain Ahab Had a Wife: New England Women and the Whalefishery, 1720–1870*, Chapel Hill, NC: University of North Carolina Press.

Phillips, A., and J. C. Sharman (2020). *Outsourcing Empire. How Company-States Made the Modern World*, Princeton, NJ: Princeton University Press.

Policante, A. (2015). *The Pirate Myth. Genealogies of an Imperial Concept*, Milton Park: Routledge.

Ray, J. L. (1989). 'The abolition of slavery and the end of international war', *International Organization* 43(3): 405–39.

Rediker, M. (1987). *Between the Devil and the Deep Blue Sea: Merchant Seamen, Pirates and the Anglo-American Maritime World, 1700–1750*, Cambridge: Cambridge University Press.

Ritchie, R. C. (1986). *Captain Kidd and the War against the Pirates*, Cambridge, MA: Harvard University Press.

Shilliam, R. (2015). *The Black Pacific: Anti-colonial Struggles and Oceanic Connections*, London: Bloomsbury Publishing.

Stanley, J. (2003). 'Women at sea: an other category', *Gender and History* 15(1): 135–9.

Steinberg, P. E. (2001). *The Social Construction of the Ocean*, Cambridge: Cambridge University Press.

Stoddard, L. (1921 [1920]). *The Rising Tide of Color against White World-Supremacy*, New York: Charles Schribner's Sons.

Stokke, O. S. (2007). 'A legal regime for the Arctic? Interplay with the Law of the Sea Convention', *Marine Policy* 31(4): 402–8.

Tenold, S., and B. G. Økland (2019). 'Mermaids ashore: the Norwegian Mermaid Association, 1964–1989', *The Mariner's Mirror* 105(2): 202–18.

Thomas, M. (2004). ' "Get yourself a proper job girlie!": recruitment, retention and women seafarers', *Maritime Policy and Management* 31(4): 309–18.

Tinsley, O. N. (2008). 'Black Atlantic, queer Atlantic: queer imaginings of the Middle Passage', *GLQ* 14(2–3): 191–215.

Towns, A. (2021). 'Gender in historical International Relations', in B. de Carvalho, J. Costa López and H. Leira, eds, *The Routledge Handbook of Historical International Relations*, Abingdon: Routledge, 153–61.

Turley, H. (1999). *Rum, Sodomy, and the Lash: Piracy, Sexuality, and Masculine Identity*, New York: New York University Press.

Wadsworth, J. E. (2019). *Global Piracy: A Documentary History of Seaborne Banditry*, London: Bloomsbury.

Waitt, G. (2008). ' "Killing waves": surfing, space and gender', *Social and Cultural Geography* 9(1): 75–94.

Windler, C. (2001). 'Diplomatic history as a field for cultural analysis: Muslim-Christian relations in Tunis, 1700–1840', *The Historical Journal* 44(1): 79–106.

Yao, J., and A. Delatolla (2021) 'Race and Historical International Relations', in B. de Carvalho, J. Costa López and H. Leira, eds, *The Routledge Handbook of Historical International Relations*, Abingdon: Routledge, 192–200.

4

Boundaries in the sea: the production of political space in the early modern colonial Atlantic

Mark Shirk

This volume reconceptualises International Relations (IR) *with* the sea. One important aspect of IR is the production of political space. Political space is produced as actors bound spaces for political purposes through practices. While not always using this language, literatures on sovereignty, state formation, and mapping in IR all examine the production of political space. However, most of this literature is exclusively about land. How can we understand the production of political space if we only look at 29 per cent of the globe's surface? Traditionally, it has been argued that the sea is an open, apolitical space used to traverse from one piece of land to another. It has unchangeable material properties that have dictated the role it plays in international politics. But the historical record says otherwise. The sea is a political space: It can be bounded, claimed, and contested.

This chapter makes the argument that understanding how political space has been produced at sea furthers our understanding of international politics. I demonstrate this by looking at shifting meanings of 'the line' as a boundary in the Atlantic from the late fifteenth to the early eighteenth century. During this period, there are three major productions of the sea as a political space that utilise the three 'tropes' that de Carvalho and Leira (2022) describe in the introduction to this volume: (1) controlling maritime resources – the sea as a repository of future discoveries and the line demarcating Spanish and Portuguese claims; (2) taming the sea – the Spanish claim the sea 'beyond the line' – now the Tropic of Cancer – but these claims are contested by rivals resulting in 'no peace beyond the line'; (3) traversing the water – the rise of England's plantation complex and the end of sea raiding as a legitimate practice push the line to the Equator where it is politically meaningless, opening the sea for trade. When utilising these tropes and placing them into historical context, we end up with three unique productions of the sea as political space.

Conceptualising the sea as a political space has a number of consequences for IR. First, scholars should think about politics and the construction of political space, not natural features of land/seascapes that have predictable effects. Second, any ocean, river, or lake has been produced as political space in multiple ways across time. Third, doing IR with the sea can tell us about the production of political space writ large. The sea and the land can tell us about each other, about the distinctions we make between them, and about processes that extend across both. I will demonstrate by highlighting how practices of violence and colonial governance were tied to different productions of the sea. Finally, as a result of producing the sea as a distinct political space, seaborne polities, and especially empires, function differently than land-based ones in ways that have been largely ignored by IR scholars.

This chapter unfolds by unpacking the production of political space as practices that draw boundaries and why that means we should be doing IR with the sea. This will be followed by a narrative on the history of the line covering each of the three productions outlined above. Due to space constraints, the narrative will be broad-brush and focus on Spain[1] and select rivals. Before concluding, the penultimate section will discuss the four aforementioned implications for IR in the context of the narrative.

The sea as political space

No physical space has inherent political meaning. How could it? The meaning of any particular space – i.e. what is 'France' vs. 'Germany', when did 'France' become a 'democracy', etc. – is not timeless. This is recognised by those studying political geography (Rose, 1993; Dalby and Tuathail, 1998; Ingram and Dodds, 2009; Agnew and Muscarà, 2012) but also in studies tracing the development of the sovereign state (Spruyt, 1994; Hall, 1999; Bukovansky,

2002; Nexon, 2009), boundaries (Goettlich, 2019), or the mapping of Europe (Branch, 2011) or empires (Bayly, 2016), among others, in IR. Since political space is produced, it can vary across time and space.

Political space is produced by boundaries. A growing literature focuses on boundaries as the locus of modern politics and the state (Lamont and Molnar, 2002; Tilly, 2004; Mayrl and Quinn, 2016; Shirk, 2019). Boundaries can be 'physical', i.e. they can be seen on a map, or conceptual, demarcating what is political or apolitical (Thomson, 1994; Jackson and Nexon, 1999), what is domestic or foreign (Walker, 1993; Leira, 2019), etc. They set the state apart from society so the former can rule the latter (Mitchell, 1991). These boundaries are drawn by practices, or habitual actions that construct actors and their conceptual maps of the world (Neumann, 2002; Adler and Pouliot, 2011). A focus on practices means a focus on observable, repeated actions as opposed to incorporeal ideas or the 'true nature' of any material substance. The practices that drew 'the line' were the product of 'many hands' (Morgan and Orloff, 2017). Early modern Atlantic empires were rooted in actors located in the outposts developing practices to help themselves solve problems (Joas, 1996) and get on with the business of agriculture, trade, religious conversion, etc. (Norton, 2017). So it is in performing the everyday, as opposed to some grand plan, that political space is often produced.

Because we can observe practices that draw boundaries in the sea we can begin to conceptualise the sea as a political space. What does this mean? First, the process of producing political space at sea is the same as it is on land – everyday practices produce boundaries that give spaces meaning. The sea does not have natural 'political' properties as traditional materialist conceptions have argued or implied (Braudel, 1996; Mahan, 2010). It is undeniable that sea and land have different properties and these properties certainly condition political possibility on each. However, while the material properties of the sea remained largely constant in the narrative below, its production as a political space was dynamic. Such an argument is newly relevant with the rise of the 'new materialisms' (Coole and Frost, 2010; Connolly, 2013) that emphasise the agency and material properties of matter and the non-human on social processes. It is unclear, however, how much the 'true' properties of the sea, in contrast to the 'wrong' properties believed to be true in any specific time and place, impact the production of political space. Therefore, how the sea is thought and used, demonstrated through meaning-making practices, matters more for *our purposes* than its material properties.

Second, this does not mean that we should ignore the sea. Materialist conceptions of the sea could lead one to think that since the sea is static, it is either unimportant to international politics (Mearsheimer, 2001) or a mere extension of land (Lambert, 2018; Till, 2018). However, there is also a danger that stating that political space is produced at sea in the same way it is produced on land could lead to us ignoring the sea. This would be a mistake that reproduces the fallacy that the sea is a naturally open, apolitical space. Doing IR with the sea means that we need to take the production of *all* political space seriously. Any study that focuses on land needs to take the sea into account because the production of political space at sea influences the same on land. The narrative below demonstrates this by looking at how the governance of colonies in the Americas changed as the sea changed. But it must be said that the inverse is also true. And the production of political space alerts us to the relational (Jackson and Nexon, 1999) processes delineating 'land' from 'sea'. The sea is also important to understanding processes that traverse the dry and the wet such as piracy and privateering. If IR scholars are to understand international politics, then we need to focus on how political space is produced across the entire globe, not just 29 per cent of it. The narrative below uses shifting placements and meanings of 'the line' in the Atlantic to make this argument.

The sea as repository

Most scholarly attention on 'the line' has focused on the placement, existence, and/or legal foundations of the line (Mattingly, 1963; Williamson, 1969; Steele, 1986; Stirk, 2011). A lot of scholarship argues that there is little evidence of 'the line' and in particular 'no peace beyond the line' in legal texts and international treaties. Garrett Mattingly finds the first use of 'no peace beyond the line' in a Walter Scott novel. He does find evidence of the line here and there and claims that it was only ever the Tropic of Cancer (Mattingly, 1963). Kenneth Steele calls 'the line' a 'whisper without an echo' (Steele, 1986). Peter Stirk argues that Carl Schmitt's focus on 'the line' was rooted in his fear of US dominance in the Americas (Stirk, 2011). Granting that the line was always contested and never set, I argue that its meaning shifted as well. The line was not drawn in a treaty but instead as a series of practices by the varied actors constituting European empires. While legal texts and treaties are the product of, and later shape, important practices, they are just one set of many and focusing instead on letters of marque, diplomatic correspondence, and patterns of sea raiding, among others, creates a picture of shifting, overlapping, and even competing meanings of 'the line'.

The line developed in the fifteenth century to demarcate spheres of discovery between Spain and Portugal. Their dispute became a common topic of papal bulls. Early bulls such as Martin V's 1418 *Sane Charissmus*, Nicholas V's 1455 *Romanus Pontifex*, and Calixtus III's 1456 *Inter Caetera* declared or reconfirmed Portuguese rights in Africa south of the Cape of Bojador (currently part of the Western Sahara). As Spain rose as a major maritime power, it began to explore the 'New World'. Castillian Pope Alexander VI gave 5 bulls recognising Spanish conquests in this sphere. His *Inter Caetera* awarded Spain everything 'one hundred leagues south and west of the Azores or Cape Verde Islands'. He brokered the 1494 Treaty of Tordesillas, which set a line 370 leagues west of the Cape Verde Islands (not specified which one). Spain claimed to the west of the line and Portugal to the East. This conveniently (for the Portuguese) cut off the, as yet undiscovered, north-eastern bulge of South America, today Brazil. The Portuguese, who at this time had a stronger navy, agreed to this split because they were preoccupied with South Asia and cared little for the 'primitive' American lands (Steinberg, 2001: 83–4).

As even contemporaries argued, papal bulls were not legally binding; the Pope did not have the authority to carve up the globe. The lines were vague and unworkable – south *and* west of a line? Which island in the Azores or Cape Verde? Such lines were frequently ignored. The two kingdoms paid no attention to papal bulls when agreeing the Treaty of Alcacoves in 1479. Neither kingdom used the bulls to justify future claims. Spain's rejection of English claims in the New World following John Cabot's mission in 1497 rested on prior discovery, not papal bulls. It wasn't until the latter half of the seventeenth century, as 'the line' was receding into meaninglessness, that Carlos II tried to use Alexander's bulls in disputes with the English (Mattingly, 1963: 153–5).

What the papal bulls, treaties, and explorations themselves tell us is that there were attempts to grant rights to land and conquest to Portugal, and later Spain, by drawing boundaries in the sea. The sea was not open but a political space containing future land acquisitions that could be claimed. This is still different to sovereign control as 'the line' was about dividing rights of movement and future conquest, not territorial control. When the Treaty of Tordesillas was concluded, Columbus had only just completed his second voyage and the continent of South America was only a rumor. Therefore, the sea was a place fecund with possible discoveries *and* one that could be drawn up into spheres of conquest (Steinberg, 2001: 82–6). The sea was similar to a piece of land with the potential to be prospected for oil or valuable minerals. However, the line and the way the North Atlantic would be produced as a political space would change.

Claiming and contesting the sea

As the battleground shifted from Africa to the West Indies, Spain claimed the sea 'beyond the line' – now basically the Tropic of Cancer – as somewhere only Spanish ships could traverse. The arrival of the French (and later the English) would make the sea a contested space. This resulted in 'no peace beyond the line' – the practice whereby 'war' beyond the line would not lead to war in Europe. While contestation adds an interesting wrinkle, Spain's rivals accepted that authority was claimed, and recognised, over sea lanes in the name of trading rights. While ocean space was not to be possessed as land (for the French and English at least), it did become a site for interstate competition and the exercise of power (Steinberg, 2011: 68–109). Even those arguing for freedom of the seas, such as Hugo Grotius, argued that states had the ability, 'to take Possession or Jurisdiction only over some Part of the Sea' (Grotius, 1962: 466). His problem was monopoly, not possession

(Stern, 2014: 183). The major threats at sea were the activities of rival states and trading lanes proved to be a legitimate *casus belli* (Glete, 2000). Interstate or inter-imperial competition was often undertaken by privateers – sea raiders with 'letters of marque' from government officials.

Shortly after arriving in the New World, Spain discovered gold and silver in South and Central America, giving it a massive advantage by controlling the raw materials for money on the continent. European rivals were unable to challenge this position directly either by capturing ports through which bullion flowed or the Caribbean Islands that could act as choke points for sea lanes. The gold mines made control of the Caribbean a necessity for the Spanish to move bullion from colony to court. For the entirety of the sixteenth century, the Spanish were the only European empire with permanent colonies in the West Indies.

By this time, France replaced Portugal as the major rival to Spanish ambitions. While France discovered and claimed areas of present-day Canada and the US Midwest, the money and power was in the Caribbean, the 'Spanish Main'. The Spanish crown attempted to enforce a monopoly on goods coming to and from its colonies; another reason why the very presence of foreign ships 'beyond the line' was illegal by Spanish law. But the colonies needed more slaves, materials, and basic goods than the Spanish crown could provide (Wright, 1929: 8–9). Spanish rivals, not recognising Spanish claims, traded – sometimes illegally, sometimes with passes – with local populations and attempted to gain access to Spanish treasure ships through sea raiding.

'No peace beyond the line' dates to this period. A common claim is that the 1559 Treaty of Cateau-Cambrésis, ending the Habsburg-Valois War between France and Spain, enshrined the idea (Williamson, 1969; Mann and Phaf-Rheinberger, 2014). However, the treaty was most concerned with continental affairs and the phrase 'no peace beyond the line' or anything similar does not appear in the text. It is more likely that, unable to find a solution to Spanish claims, the two rivals simply agreed to continue their rivalry in the West Indies through a handshake agreement that left no trail and did not include the English (Mattingly, 1963). However, by the 1598 Treaty of Vavrin, there is talk of 'lines of amity', so there was some recognition between the French and Spanish (Bridenbaugh and Bridenbaugh, 1972).

The Spanish believed their claims gave them the right to search and attack any foreign ship that entered their space. Not recognising Spanish claims, the French began to authorise sea rovers to attack Spanish ships in retaliation, but only 'beyond the line' (Mattingly, 1963). While extraction made it necessary to control the Caribbean, it also gave France, and later England, a place to attack. Prior to 1559, France and Spain were usually at war and after 1559 many French privateers were Huguenots not authorised by the crown, all of which made devising a law difficult. However, no declaration of war resulted from such attacks. The French were claiming as late as 1614 that in the West Indies 'whoever proves stronger shall be taken for the lord' (Steele, 1986: 190). The French response was not recognition of Spanish rights but an acceptance of their production of the sea, resulting in a continued sphere of war separate from 'Europe'.

By the end of the sixteenth century, England becomes Spain's major rival in the region. With no permanent settlements, England's actions in the sixteenth-century Colonial Atlantic were mostly about discovery, trade, and gaining access to Spanish gold and markets. The English were receiving licences to trade in the region from the Spanish in the 1560s.[2] During the Habsburg-Valois War, Spain used England as a counter to the French. After the war, England began to assert its interests in the region. In 1568, fighting broke out in San Juan de Ulloa between English trader John Hawkins and a Spanish fleet claiming he had no right to be in the New World. The English claimed Spanish gold ships that docked at Portsmouth. Seven Protestant provinces of the Netherlands revolted against Spain with English sympathies. These events changed the outlook between Spain and England and ultimately led to the 1573 Convention of Nijmegan, where the English agreed to end support for West Indian raiding by the likes of Hawkins and Frances Drake.

Until 1593, Elizabeth had no official knowledge of the French understanding about lines of amity. England had been at war with Spain and would be for the rest of her reign. 'No Peace' is only visible in peacetime and she had no occasion to take any notice of it. However, James I received a pair of letters from the French Queen

Regent Marie de Medici that claims that no one recognises the 'King of Spain as King and Lord of the Indies' and that 'beyond the line ... there has never been any kind of peace' between France and Spain (Mattingly, 1963: 149). The Providence Island Company requested in 1635 that they be able to 'defend themselves and to offend if they be offended, being beyond the line' (Steele, 1986: 191). Oliver Cromwell quipped, 'everyone could act for his own advantage in those quarters' (Davenport, 2012: 42). Thus, England joined France in challenging Spanish claims and accepting Spain's production of the sea – a set of throughways to be claimed and contested.

It was the English who were able to first settle colonies 'beyond the line'. Their first settlement was St Christopher (St Kitts today) in 1624 and they added Barbados (1627), Nevis (1628), and Antigua (1630) within a decade. The French had aims at St Christopher and Nevis but were driven off the islands by the English (England and France were as much rivals with each other as Spain) before settling Martinique in 1635. By this time, Spain's power on the continent was ebbing and its navy was not considered better than its rivals. However, neither the French nor the English were able to take the port cities through which bullion flowed due in large part to diseases like yellow fever and malaria brought to the region with slave ships (McNeill, 2010). However, they were able to force Spain into trading concessions 'beyond the line'. The 1648 Treaty of Munster gave the Dutch rights to trade in the Caribbean and the 1667 Treaty of Breda gave the English rights (Steele, 1986: 191). The Treaty of Utrecht in 1714 awarded England the asiento, a monopoly on slave trading to Spanish colonies (Scarlett, 1939). This demonstrates a de facto recognition that Spain 'controlled' the Caribbean; it was giving concessions, even if its grip was loosening.

Producing the sea as an area to be traversed yet claimed had an impact on how colonies were governed and violence was used. Both demonstrate how political processes are co-constitutive with how the sea is produced as political space. While not independent, colonial governors were given wide latitude of action over non-trade policy, courts, native relations, and privateering commissions. The important function wasn't control over daily governance but making sure that gold and silver flowed into Spanish coffers. This flowed from, and simultaneously allowed for, the use of resources to protect sea lanes claimed beyond the line. The line even patterned trade behaviour as joint stock companies only developed for trade on the other side of the line, not with other Europeans.

As mentioned above, 'no peace' took the form of sea raiding and it was conditioned by the line. For instance, Henry Morgan's attack on Porto Bello in 1668 brought an 'order to Spanish West Indian governors to proclaim war against the English *only* beyond the Tropic of Cancer' (Steele, 1986: 191). Neither Spain nor England stuck to the Convention of Nijmegen. Spain still claimed the Caribbean as its own and saw as its right the ability to grant access to other European traders. As a result, the Spanish handed out letters of marque to 'cruise in the Caribbean, looking for "unlicensed traders"' (Mattingly, 1963: 156) after peace treaties in 1598 (Vervins, with France), 1604 (London, with England), 1630 (Madrid, with England), 1659 (Pyrenees, with France), and 1670 (Madrid, with England). The English and French would retaliate by distributing their own letters of marque.

However, 'the line' made it easier for sea raiders to ask forgiveness rather than permission, blurring the line between pirate and privateer (Benton, 2010: 113). In 1671, Morgan strayed beyond his letter to sack the Spanish-controlled Panama City following a peace agreement between England and Spain. One observer commented that Morgan and his men 'committed many ... cruelties. They showed little mercy, even to the monks ... Nor did they spare the women, except for those who yielded themselves completely' (Exquemelin, 1969: 200–1). Morgan was arrested and sent to London but instead of punishment, King Charles II knighted Morgan and appointed him lieutenant governor of Jamaica (Exquemelin, 1969: 186–201). Government officials thought Morgan barbaric but they deemed his services beneficial, a part of the broader colonial political economy in which the sea was produced as a contested, claimable area. In fact, privateers like Morgan or Francis Drake or John Hawkins were lauded as patriotic heroes.

The practices constructing 'no peace beyond the line' allowed for competition without risking war. The English and French gained access to Spanish gold and markets, while the Spanish were allowed to seize 'illegal traders'. No empire liked the arrangement, but neither was it so bad as to force change. While 'the line' eventually gets a cut-out for North America, English domination of the region and its lack of gold meant the Caribbean was still the major battle-ground. During this time the line was a combination of two things: a) Spain claiming the sea south of the Tropic of Cancer and b) a recognition that the sea at some imaginary line layered sovereignty into two distinct legal and political systems – Europe/state and the New World/Empire. Politics was profoundly shaped by claiming use of, and drawing boundaries in, the sea. As we get to the end of the seventeenth century this begins to fall apart, with consequences for the future of sea raiding and Atlantic empires.

Opening the sea

By the middle of the eighteenth century the line was pushed so far south, to the equator, that it was no longer a relevant part of regional politics. Many saw the line as demarcating the passage to the Pacific, too far from Europe to be political controllable (Mattingly, 1963). There was no longer 'no peace beyond the line'. Violence in the colonies could theoretically cause war on the continent as it did with English entry into the War of Austrian Succession.[3] Additionally, peacetime privateering was no longer an acceptable practice, resulting in the 'golden age of piracy'. This happened for a set of interrelated reasons.

First, Spain decreased in power, especially on the continent, while England took its place as the dominant power in the Atlantic as demonstrated by the trading concessions discussed above. Second, the political economy of the Atlantic shifted. England developed its plantation colonies based around the trade of slaves and agricultural materials. They proved so successful that taxation of trade became more lucrative than the extraction of precious metals. North American tobacco and cotton and West Indian sugar replaced Central and South American bullion as goods of choice. These goods depended on West African slaves, resulting in the infamous triangular trade. Trade functioned best in peace and it became harder to justify continued warfare beyond the line. It is no coincidence that 'the line' was abandoned in English letters by the 1670s (Mattingly, 1963: 160–1).

Third, the new political economy transformed the landscape for sea raiding (Norton, 2014: 1537–75). Attempts to enforce peace beyond the line started in the middle of the seventeenth century with the allowance of a grace period of nine months after signing a peace treaty – which was ignored in Morgan's case. However, it was not until the 1670s, when the line was no longer mentioned by the English, that England began to tell colonial governors that they could not undertake measures of war without London's blessing. In 1689, all West Indian governors were barred from issuing letters of marque (Labaree, 1967: doc. 43). Nonetheless, the practice continued. Jamaican Governor Archibald Hamilton gave out letters of marque to privateers under the guise of fighting piracy well into the eighteenth century (TNA ADM 1/1471; CO 137/11).

The fates of sea raiders between the Nine Years War and the War of Spanish Succession (roughly 1697–1703), the 'Red Sea Men', demonstrate this. Henry Every (Burgess, 2009), William Kidd, Thomas Tew, and John Quelch all set sail with legitimate letters of marque; all acted as sea raiders, like Henry Morgan, had for decades. Every became the most wanted man in two oceans, Kidd was hanged for taking a ship covered in his commission, Tew was attacked at sea and fatally shot, Quelch hanged due to a peace between England and Portugal that did not exist when he took sail. In 1699, the English government passed the Act for the More Effectual Supression of Piracy, labelling pirates *hostis humani generis* or enemies of humanity (Rubin, 1988).

European empires would find themselves back at war in 1703, stalling any progress on sea raiding. But following the Treaty of Utrecht in 1714, pirates claiming to be 'villains of all nations' terrorised trade. A confluence of factors produced nearly perfect working conditions for pirates. As Marcus Rediker puts it, 'The sailor knew ... the Atlantic was a big place, that the empires were overstretched ... these circumstances created openings from below' (Rediker, 2004: 24). There was still demand for stolen goods and many colonists, depending on cheap goods, were favourable to sea raiders. Sea raiding transformed the economy of Bermuda (Jarvis, 2012: 84–5). One English official remarked that, 'The Pirates themselves have often told me that if they had not been supported from the traders from thence [New York, Pennsylvania, and

Rhode Island] with ammunition and provisions according to their directions, they would never become so formidable' (TNA CO 23/1:f.47). As England's interests changed to developing a trade-based economy in the Atlantic, the openness of colonial markets to pirate goods became a problem.

The pirates of the golden age would embrace *hostis humani generis* (TNA CO 23/1:f.76) to cause a deep crisis for Atlantic empires. Roughly 2,400 English ships were taken in the years 1716–26, more than the privateering ventures of any state during the just completed war. In 1720 alone, English merchants lost seventy slave missions, each valued at £3,000 (or £69,560,000 in 2020 value as a share of GDP).[4] Virginia and Maryland merchants complained that in 1717 pirates had cost them £300,000 (TNA SP 42/123). Commerce into and out of ports such as Charleston and Philadelphia were interrupted for days or even weeks at a time. In addition, the legitmacy of both state and empire were at stake. As Matthew Norton argues, state power 'depended on the ability of state agents to engage in an extended, coordinated pattern of performances that construed piracy as a distinctive social object, defined by violent state repression' (Norton, 2014: 1572).

Despite some success,[5] naval power was largely a failure in the fight against piracy since naval ships tended to be quite cumbersome and unable to chase pirates into shallower waters (TNA CO 388/24:f.186). Pardons also failed as pirates such as Blackbeard and Charles Vane accepted them and went right back 'upon the account'.[6] What did work were actions that ended any vestiges of 'the line'. The proliferation of Vice Admiralty courts into the colonies (Benton, 2010: 145–8) proved effective in prosecuting pirates[7] and working around sympathetic colonial populations (Ritchie, 1986: 144). This standardised piracy law across the empire[8] and negated the choice between a friendly jury in the colonies or a long journey to London. England replaced colonial governors in New York,[9] Jamaica,[10] and the Bahamas,[11] among others, who did not follow the line on piracy. The same thing held in French and Spanish colonies as well.[12] A propaganda campaign developed that depicted pirates as 'instigated by the devil' and 'Devils incarnate' (Rediker, 2004: 152–3), rhetoric the pirates themselves echoed. All three strategies brought Atlantic colonies closer to the state, further eroding 'the line'. This in turn would put pressure on England's

North American empire, creating the structural conditions for the revolt that would follow a half century later. Pirates still existed in the Colonial Atlantic after 1726, but they were rare and easier to deal with.[13] Piracy would not became a major problem in the region until the early nineteenth century and then for different reasons with different consequences (Earle, 2003: 209–28; Benton, 2011: 225–40).

To sum up, by the end of the seventeenth century England was beginning to move away from 'no peace beyond the line' due to its growing interests in trade and plantations. The way it dealt with Spain was to reject Spain's production of the sea as claimable, preferring to open it instead. However, 'the line' still impacted colonial governance, creating perfect working conditions for piracy. Finding solutions to the 'golden age of piracy' meant bringing Atlantic colonies closer to the metropole. In other words, the sea could only be open for trade if boundaries were drawn so that pirates could be controlled. This pushed the line south to the equator, both ending the significance of the line and enabling future problems for Atlantic empires.

Producing the sea in the early modern colonial Atlantic

In the introduction, I provided four implications of this argument for doing IR with the sea. In this section, I want to review those implications considering the historical narrative of 'the line' in the early modern colonial Atlantic. The sea is not a naturally open space. The North Atlantic of the mid-eighteenth century may resemble the ideal of an open ocean. Compared to previous centuries, they were violence-free during peacetime, trade moved across it with little friction, and fewer shipping lanes were claimed. However, Spanish claims on the Caribbean show an alternative construction of ocean space as political space. French and English attempts to be active in the region show that the sea can be contested, which would be impossible for an 'open sea'. And the first signs of an open North Atlantic happened as a result of politics, not the sea imposing its natural order. Of course, I can only claim that this story shows the opening of the North Atlantic; other bodies of sea will have different historical trajectories. The sea was not magically open by 1750; it was opening with emphasis on the process. And I am not claiming that the North Atlantic of 1750 is the same as the North Atlantic of 2022. Much has changed, too much to document in this space. But those changes do tell us that an 'open' sea has many variations, once again pointing to the importance of politics in constructing the sea as opposed to the natural properties of the sea.

It would also be wrong to suggest that this is a story of the Spanish 'getting it wrong' and then finally learning as the English gained an upper hand by understanding the sea's true nature. Nor is it the story of any empire 'taming' the sea. Spanish claims to the Caribbean lasted from the late fifteenth century well into the eighteenth. That is roughly as long as there has been an open sea in the North Atlantic. And it worked for them! It would be more accurate to say that the Spanish produced a politics of the sea that was different from the norm today and it lasted for centuries. It may have lasted longer – in its contested form, as Spanish hegemony may have been shorter – had yellow fever or the plantation complex not been present. And it finally changed as local actors developed new practices to work through local problems of sea raiding, trade, disease, and colonial governance. There is no reason to think that the way we construct the sea as political space today is the inevitable end point; other systems have been and will be possible going forward.

In the spirit of IR with the sea, we need to study the sea because it will give us a better conception of political space. The narrative above provides two examples. First, in the Spanish production of the sea, colonies were given wide latitude to do what they will so long as they were able to provide raw materials. In what we might call the English production, colonies had to be controlled to make sure that the sea was an open space for trade. In this shift we can see how land (the development of plantations) helped to reconstruct the sea, while the sea (opening for trade) helped to reconstruct the political situation on land. What we see is relational – land and sea constructed by the transactions of actors inhabiting and using each. This isn't about knowing land per se but instead understanding how politics and political space have been constructed across time. Second, sea raiding went from a legitimate part of interstate competition to a practice that put its purveyors outside of civilisation. This stands as one of the first instances of private violence in the West (Mabee, 2009). A practice that takes place at sea but has implications on land (sea raiding), and a broader category of violence that takes place on both (private violence), can only be understood by taking the production of the sea as a political space seriously. Without studying the sea, we cannot fully comprehend international politics.

Finally, a note on seaborne empires. This is another datapoint to go with studies emphasising the diversity and difference of seaborne and land-based empires (Phillips and Sharman, 2015; Sharman, 2019). It is little coincidence that those European states – England and the Netherlands – most committed to different conceptions of an 'open sea' were also those most successful at building long-lasting maritime empires in the early modern era. This is because their seaborne empires functioned differently than land-based empires in both Europe and Asia and the rival Spanish seaborne empire. Land-based empires were based on military might, seaborne empires on open seas were built on networks of commerce and exchange. This is why Jason Sharman has labelled them 'Empires of the Weak' – the start-up capital to build and maintain them was less than the Mughal, Ottoman, or Holy Roman empires because they were as much about connections – traversing the water – as war and rule.

Conclusion

The point of this chapter is not to get 'the line's' placement right or prove its legal existence. Nor is it about the line becoming the equator, causes of the American Revolution, or the end of piracy. It is not the destination that is important but the journey. What we can learn from the journey is that the sea – in this case the Atlantic Ocean and the Caribbean Sea – is a political space. It can, and has, been bounded. 'The line' is evidence of that. Its meanings have changed as a result of bounding practices and contestation. 'The line' was first drawn to divvy up rights to discovery between Spain and Portugal in the late fifteenth century. Hence the sea was a repository for future discoveries. By the middle of the sixteenth century, Spain discovered gold and silver mines and claimed not just most of present-day Central and South America and the West Indies but also the sea that connected them. The Spanish made the presence of foreign ships 'beyond the line' illegal. The French balked and refused to concede the Caribbean Sea to the Spanish, creating the conditions for 'no peace beyond the line'. The sea was a claimed and contested space; the line separated the colonies from Europe, layering sovereignty. The English found another way to power in the region and in the process challenging Spain and combatting piracy pushed the line so far south as to make it politically meaningless, essentially opening the sea.

The 'open sea' is one of many possible constructions, not a natural condition. And it is only open due to new boundaries drawn to make it that way. I want to make it clear that I am not claiming that the North Atlantic, let alone the entire ocean, was fully open starting in 1730 or so and has remained so today. That is clearly not true. Jeppe Mulich's 'microregionalism' (2013) is just one of many variants that took place in this region. Instead, we should think of this as opening the sea, with the emphasis on process and beginning. Second, an open sea can take many forms; clearly the 'open sea' of today is much different than that of the eighteenth century. To do IR with the sea, further research needs to be done tracing political developments on sea in other times and places, focusing on its dynamism and the ways that it has or has not interacted with developments on land. In this chapter I have attempted to demonstrate that the sea is a political space whose production has changed over time and these changes have impacted multiple political processes of interest to IR. For this reason, we need to bring the sea into our studies of war, trade, cooperation, etc.; do IR with sea.

Notes

1 The 'Spanish' empire was constituted of polities that could be called Castile, the Holy Roman Empire, and Spain, among others. Following convention, the term 'Spain' is used here. Ditto for their rivals.
2 For licences given to John Hawkins to sail in Spanish waters and trade at Spanish ports, see Wright (1929: doc. 4, 14).
3 The English justified declaring war on Spain in 1739 as retaliation for a 1731 attack on merchant Robert Jenkins. It became known as 'The War of Jenkins' Ear', though it was probably fear of losing the lucrative asiento trading rights that brought England into war.

4 The RPI value is £400,000, the labour value £5–6 million (Officer and Williamson, 2015).
5 A naval mission that attacked and killed the prolific Bartholomew Roberts (TNA ADM 2/50: f.290–3).
6 See CSPCS 31/33; 32/527; 30/738; 32/256:f.166; TNA ADM 51/690; CO 5/867:f.7; 23/13:f.28–9.
7 There were 418 documented hangings of pirates from 1716 to 1726 (Frykman, 2014: 224).
8 This continued to be a problem into the 1720s; see CSPCS 9/295; TNA 1/1879:9 Jul 1718; CO 5/1364:f.505–520; 138/3:f.81–3; 323/3:f.298; 323/2:f.271–6; 323/8:f.10; 324/6:f.163.
9 TNA CO 5/1116:f.2–12; 391/8:f.10.
10 TNA CO 137/11:f.92–109; 137/12:f.103.
11 CSPCS 29/635:f.388; 30/556:f.266; TNA CO 5/1317: f.245–5; 5/1364:f.483–7; 23/12:f.103–4, 107; 152/12/1:f.34; SP 42/16:5 Aug 1716.
12 TNA CO 388/24:f.184.
13 For proceedings of a trial for Piracy in 1740, see TNA HCA 1/99.

References

Adler, E., and V. Pouliot (2011). 'International practices', *International Theory* 3(1): 1–36.
Agnew, J., and L. Muscarà (2012). *Making Political Geography*, London: Rowman and Littlefield.
Bayly, M. J. (2016). *Taming the Imperial Imagination: Colonial Knowledge, International Relations, and the Anglo-Afghan Encounter, 1808–1878*, Cambridge: Cambridge University Press.
Benton, L. (2010). *A Search for Sovereignty*, Cambridge: Cambridge University Press.
Benton, L. (2011). 'Toward a new legal history of piracy: maritime legalities and the myth of universal jurisdiction', *International Journal of Maritime History* 23(1): 225–40.
Branch, J. (2011). 'Mapping the sovereign state: technology, authority, and systemic change', *International Organization* 65(1): 1–36.
Braudel, F. (1996). *The Mediterranean and the Mediterranean World in the Age of Philip II*, vol. 1, Berkeley, CA: University of California Press.
Bridenbaugh, C., and R. Bridenbaugh (1972). *No Peace Beyond the Line: The English in the Caribbean, 1624–1690*, Oxford: Oxford University Press.

Bukovansky, M. (2002). *Legitimacy and Power Politics: The American and French Revolutions in International Political Culture*, Princeton, NJ: Princeton University Press.

Burgess, D. R. (2009). 'Piracy in the public sphere: the Henry Every trials and the battle for meaning in seventeenth-century print culture', *Journal of British Studies* 48(4): 887–913.

Connolly, W. E. (2013). 'The "New Materialism" and the fragility of things', *Millennium – Journal of International Studies* 41(3): 399–412.

de Carvalho, B., and H. Leira (2022). 'Introduction: staring at the sea', in B. de Carvalho and H. Leira, eds, *The Sea and International Relations*, Manchester: Manchester University Press.

Coole, D., and S. Frost, eds (2010). *The New Materialisms: Ontology, Agency, and Politics*, Durham, NC: Duke University Press.

Dalby, S., and G. O. Tuathail, eds (1998). *Rethinking Geopolitics*, London: Routledge.

Davenport, F. G. (2012). *European Treaties Bearing on the History of the United States and Its Dependencies*, Clark, NJ: The Lawbook Exchange.

Earle, P. (2003). *The Pirate Wars*, New York: St Martin's Press.

Exquemelin, A. O. (1969). *The Buccaneers of America*, trans. A. Brown, Mineola, NY: Dover Publications.

Frykman, N. (2014). 'Pirates and smugglers: political economy in the Red Atlantic', in P. J. Stern and C. Wennerlind, eds, *Mercantilism Reimagined: Political Economy in Early Modern Britain and Its Empire*, Oxford: Oxford University Press, 218–39.

Glete, J. (2000). *Warfare at Sea, 1500–1650: Maritime Conflicts and the Transformation of Europe*, London: Routledge.

Goettlich, K. (2019). 'The rise of linear borders in world politics', *European Journal of International Relations* 25(1): 203–28.

Grotius, H. (1962). *The Law of War and Peace*. De Jure Belli Ac Pacis, Libri Tres, The Essay and Monograph Series of the Liberal Arts Press. Indianapolis, IN: Bobbs-Merrill.

Hall, R. B. (1999). *National Collective Identity: Social Constructs and International Systems*, New York: Columbia University Press.

Ingram, A., and K. Dodds, eds (2009). *Spaces of Security and Insecurity: Geographies of the War on Terror*, London: Routledge.

Jackson, P. T. and D. H. Nexon (1999). 'Relations before states: substance, process, and the study of world politics', *European Journal of International Relations* 5(3): 291–332.

Jarvis, M. J. (2012). *In the Eye of All Trade: Bermuda, Bermudians, and the Maritime Atlantic World, 1680–1783*. Chapel Hill, NC: University of North Carolina Press.

Joas, H. (1996). *The Creativity of Action*, Chicago: University of Chicago Press.

Labaree, L. W., ed. (1967). *Royal Instructions to British Colonial Governors*, New York: Octagon Books.

Lambert, A. (2018). *Seapower States: Maritime Culture, Continental Empires, and the Conflict That Made the Modern World*, New Haven, CT: Yale University Press.

Lamont, M., and V. Molnar (2002). 'The study of boundaries in the social sciences', *Annual Review of Sociology* 28: 167–95.

Leira, H. (2019). 'The emergence of foreign policy', *International Studies Quarterly* 63(1): 187–98.

Mabee, B. (2009). 'Pirates, privateers, and the political economy of private violence', *Global Change, Peace, and Security* 21(2): 139–52.

Mahan, A. T. (2010). *The Influence of Sea Power upon History, 1660–1783*. Cambridge: Cambridge University Press.

Mann, M., and I. Phaf-Rheinberger, eds (2014). *Beyond the Line: Cultural Narratives of The Southern Oceans*, Berlin: Neoflis Verlag UG.

Mattingly, G. (1963). 'No peace beyond what line?' *Transactions of the Royal Historical Society* 5(13): 145–62.

Mayrl, D., and S. Quinn (2016). 'Defining the state from within boundaries, schemas, and associational policymaking', *Sociological Theory* 34(1): 1–26.

McNeill, J. R. (2010). *Mosquito Empires: Ecology and War in the Greater Caribbean, 1620–1914*, Cambridge: Cambridge University Press.

Mearsheimer, J. J. (2001). *The Tragedy of Great Power Politics*, New York: Norton.

Mitchell, T. (1991). 'The limits of the state: beyond statist approaches and their critics', *American Political Science Review* 85(1): 77–96.

Morgan, K. J., and A. S. Orloff, eds (2017). *The Many Hands of the State: Theorizing Political Authority and Social Control*, Cambridge: Cambridge University Press.

Mulich, J. (2013). 'Microregionalism and intercolonial relations: the case of the Danish West Indies, 1730–1830', *Journal of Global History* 8(1): 72–94.

Neumann, I. B. (2002). 'Returning practice to the linguistic turn: the case of diplomacy', *Millennium – Journal of International Studies* 31(3): 627–51.

Nexon, D. H. (2009). *The Struggle for Power in Early Modern Europe: Religious Conflict, Dynastic Empires, and International Change,* Princeton, NJ: Princeton University Press.

Norton, M. (2014). 'Classification and coercion: the destruction of piracy in the English maritime system', *American Journal of Sociology* 119(6): 1537–75.

Norton, M. (2017). 'Real mythic histories: circulatory networks and state-centrism', in J. Go and G. Lawson, eds, *Global Historical Sociology*, Cambridge: Cambridge University Press.

Officer, L. H., and S. H. Williamson (2015). 'Five ways to compute the relative value of the UK pound amount, 1270 to present', *Measuring Worth*. www.measuringworth.com/calculators/ukcompare (accessed 14 June 2021).

Phillips, A., and J. C. Sharman (2015). *International Order in Diversity: War, Trade and Rule in the Indian Ocean*, Cambridge: Cambridge University Press.

Rediker, M. (2004). *Villains of All Nations: Atlantic Pirates in the Golden Age*, Boston, MA: Beacon Pres.

Ritchie, R. C. (1986). *Captain Kidd and the War Against the Pirates*, Cambridge, MA: Harvard University Press.

Rose, G. (1993). *Feminism and Geography: The Limits of Geographical Knowledge*, Minneapolis, MN: University of Minnesota Press.

Rubin, A. P. (1988). *The Law of Piracy*, Newport: Naval War College Press.

Scarlett, G. C. (1939). *The Treaty of Utrecht*, New York: Cosmos.

Sharman, J. C. (2019). *Empires of the Weak: The Real Story of European Expansion and the Creation of the New World Order*, Princeton, NJ: Princeton University Press.

Shirk, M. A. (2019). 'The universal eye: anarchist "propaganda of the deed" and development of the modern surveillance state', *International Studies Quarterly* 63(2): 334–45.

Spruyt, H. (1994). *The Sovereign State and Its Competitors: An Analysis of Systems Change*, Princeton, NJ: Princeton University Press.

Steele, I. K. (1986). *The English Atlantic, 1675–1740: An Exploration of Communication and Community*, Oxford: Oxford University Press.

Steinberg, P. E. (2001). *The Social Construction of the Ocean*, Cambridge: Cambridge University Press.

Stern, P. J. (2014). 'Companies: monopoly, sovereignty, and the East Indies', in P. J. Stern and C. Wennerlind, eds, *Mercantilism Reimagined: Political Economy in Early Modern Britain and Its Empire*, Oxford: Oxford University Press, 177–95.

Stirk, P. (2011). 'No peace beyond the line', in S. Legg, ed., *Spatiality, Sovereignty, and Carl Schmitt: Geographies of the Nomos*, London: Routledge, 276–83.

Thomson, J. (1994). 'State sovereignty in international relations', *International Studies Quarterly* 39(2): 213–33.

Till, G. (2018). *Seapower: A Guide for the 21st Century*, London: Routledge.

Tilly, C. (2004). 'Social boundary mechanisms', *Philosophy of the Social Sciences* 34(2): 211–36.

Walker, R. B. J. (1993). *Inside/Outside: International Relations as Political Theory*, Cambridge: Cambridge University Press.

Williamson, J. (1969). *Hawkins of Plymouth*, New York: Barnes and Noble.

Wright, I. A., ed. (1929). *Spanish Documents Concerning English Voyages to the Caribbean, 1527–1568*, London: Hakluyt Society.

5

Challenging order at sea: the early practice of privateering

Benjamin de Carvalho and Halvard Leira

By the time most European polities started looking overseas in
earnest in the late fifteenth century, the Iberian powers had already
been able to secure papal sanction for what amounted to a global
duopoly.[1] The Treaty of Tordesillas gave Spain and Portugal exclusive rights to half of the world each. A century later, both the duopoly
and the religious order of Europe had been upended. A key practice in this upending was that of privateering. Privateering played
a crucial part both in the survival of Protestantism in Europe and
in the spread of the European-dominated state system, accounting
largely for how polities beyond the Iberian ones went overseas
and how they came to settle around the world. By continuously
pushing boundaries, privateering proved to be a perfectly suited
practice for challenging the Iberian imperial duopoly and successfully transforming it into a wider European imperial venture.
Understanding privateering, then, opens the door to making sense
of the challenge posed by the sea to different European polities,
how they managed to overcome the obstacles posed by the sea, and
how the sea became a political fibre, structuring the reach of their
political authority. By challenging traditional dichotomies of public
and private, sea and land, state and empire, trade and war, privateering is a clear-cut example of that rethinking we called for in the
introduction, a rethinking of international relations with the sea.

In this chapter we explore the early practice of privateering
along two axes. First, Protestant privateering was central to the
survival of Protestantism in north-western Europe, in particular in
a first phase from around 1560 to the late 1580s; second, and just
as importantly, Protestant privateering had global repercussions
by breaking the Iberian imperial duopoly, establishing new polities

overseas and enabling and necessitating a more robust imperial presence. Where Protestant privateering in the first phase was largely parasitic (see Löwenheim, 2007), in the second phase it was transformative, changing both the European colonial projects and the legal underpinnings of the whole system, as seen most explicitly in the work of Grotius, by completely overturning the social construction of the ocean (see Steinberg, 2001). Put simply, Protestant privateering deeply influenced the trajectory of global politics for centuries.

We approach the topic in four steps. Starting with a brief overview of what privateering consisted of and how it was practised and regulated, we then discuss the continental context of confessional divides and how they impacted the policies of Protestant states. The main part of the chapter is then concerned with the three cases of Protestant privateering, Huguenot, English and Dutch.[2]

Privateering: historical trajectory, practice and regulation

The practice of privateering emerged gradually in the late Middle Ages, likely as a result of increasing trade between people owing allegiance to different overlords. With no system of law in place for regulation of interaction between people living under different jurisdictions, privateering seems to have developed as a sort of tit-for-tat self-help system: 'Property taken by or outstanding debts against foreigners could be recovered, if judicial recourse was not available, by capturing goods from foreigners of the same nationality' (Thomas, 2003: 364). Eventually, the acknowledged 'right' to act in this way became tied to the issuing of a letter of marque or reprisal by a sovereign. From the personalised roots, the practice gradually evolved to a general one, allowing anyone to apply for a privateering commission against the enemy in wartime. To organise privateering, states established prize courts in home ports and consular posts abroad (Leira and Neumann, 2008; Leira and de Carvalho, 2011) and took considerable care to have fair adjudication before a prize could be sold for profit. Much literature lazily conflates privateering and piracy, and there was indeed considerable overlap in personnel and similarities in practice, but the legal framework and the state involvement nevertheless entailed that privateering must be understood as a separate and specific practice. And, we will argue, as a much more important one.

A common view of privateering has been that it was the weapon of the weak – used by polities which were unable to raise a regular navy, for lack of manpower or finances. While this notion is not completely unfounded, it is certainly too simplistic. For some states generally considered weak in this respect, speed seems to have been just as important as costs. A privateer navy could be raised more or less overnight, by issuing a number of letters of marque or reprisal, utilising already existing ships and sailors, while a regular navy took time to build and man. Yet other states turned to privateering as a matter of strategy. Stressing 'weak states' furthermore completely ignores the other part of the equation – the providers of the means of violence. While certainly a tool for the state (or other polities) during much of its existence, privateering was in its original form a 'private' undertaking, where one sought to redress wrongs committed by individuals subordinate to one legitimate authority against individuals subordinate to another legitimate authority. The polities were primarily providing sanction for this undertaking. Increasingly, privateering became a more general means for enrichment for investors and outfitters, and a means for survival for mariners of all stripes. Thus, in many cases, privateering was a weapon not so much of weak polities, as of thrifty merchants and hungry sailors. Finally, privateering was also at times part of the ideological struggles over the construction of the ocean. Could the oceans be divided in the same way as territory, sanctioned by religious credo, or were the oceans in principle free? Could neutrals carry goods to belligerents, or were such goods generally contraband? When challenging existing notions of legality and legitimate authority at sea, states would often turn to privateers. By shadowing into piracy, privateering enabled plausible deniability by the challenging states, while the very act of issuing privateering commissions undercut claims to supreme political and/or religious authority.

Privateering would persist until outlawed by the Treaty of Paris in 1856, but the practice was transformed around 1600, as part and parcel of the wider transformation in the thinking about, and practice of, war and peace. The main protagonists in this transformation were the Protestant privateers.

The continental background of overseas predation:
the Protestant cause

While privateering had been a constant of European waters since at least the twelfth century, the fervour with which it was pursued around 1600 was newfound. It seems unlikely that this increase in scale and scope was not related to the religious conflicts tearing at the fabric of Christendom. The second half of the sixteenth century was a period of intense religious competition and conflict in western Europe. The peace of Augsburg had brought a measure of peace to the empire, but west of that, religious wars flared. France went through an almost continuous period of civil war from 1562 to 1598, the Dutch rebelled against Spain in 1568, leading to a war which would not end permanently until 1648, and England was dealing uneasily with Spain from the 1560s, with tension erupting into full-scale war from 1585–1604. The importance of religious (in particular, Protestant) ties during this period can hardly be overstated. While states and political leaders were certainly looking out for their own interests, those could not necessarily be detached from their religious beliefs and identities.

For most of the period under consideration here, England was the key to Protestant hopes and alliances (see Sutherland, 1992: 596–600). At her accession in 1558, Elizabeth inherited a war between France and the Catholic alliance of England and Spain – an alliance whose limits were to be toyed with until war between Spain and England broke out openly. It makes good sense to interpret England's foreign policies from Elizabeth's ascension in 1558 as marked by constancy and ideological direction grounded in England's identity as a Protestant state (see de Carvalho, 2014). England's pose was guided by the need to balance between what was possible to achieve, given England's relatively weak military position, against the strong commitment to Protestantism at home and abroad.

For our purposes, the solidarity among co-religionists, the simultaneity of conflict and the location of all Protestant strongholds on the North Atlantic seaboard had a number of important consequences. At an overall level, the common Protestant cause often led to political, economic and military support. More specifically, the four following points stand out. First, there was a

possibility for joint action, with privateers of different nationalities working together against common enemies. Second, there was
virtually always a friendly port for privateers to seek out. Third,
and closely related, it would more often than not be possible to
secure privateering commissions abroad, if the home polity for one
reason or another was unwilling to grant them. Fourth, solidarity
eased migration. Numerous Huguenots migrated from France to
England, as did 'flocs of heretics' from the Netherlands according
to the Spanish ambassador La Quadra (quoted in Trim, 2005: 148),
and likewise from the southern part of the Low Countries to the
northern one. Such migrations placed exiles with a lust for revenge
in friendly places, further fuelling the privateering campaigns
against the Catholic enemies.

Protestant military engagements on the continent were risky,
and the repercussions potentially disastrous for Protestant powers
which still could not match the sheer military might of Catholic
powers. Engagement at sea, on the other hand, was a different
matter, and could be conducted through means which were legitimate within an old legal order, yet could provide the instruments to
challenge or change that same order. Privateering, sitting uneasily
between medieval conceptions of warfare and the new order, and
between state action and private individuals, provided the means
for such an engagement. Accordingly, over the last three decades of
the sixteenth century, the enmity between Protestant and Catholic
powers came to turn towards the seas and the ocean. This move
came to be of lasting importance for the further development of the
international system. While Protestant polities had been concerned
mainly with securing their independence from Catholic challengers
up until the 1570s, as the era of Protestant privateering came to
an end half a century later, Protestant power had been projected
globally, and was redefining Catholic spaces all around the world.

Huguenot privateering

The development of privateering in its traditional sense is inherently linked to the early wars of religion in France, and notably
to the role played by the Huguenot leader and Admiral of France
Gaspard de Coligny – who would later be assassinated in the
St Bartholomew's Day Massacre in 1572. Yet, the birth pangs of

French privateering are to be found much earlier. Invigorated by Francis I's refusal to acknowledge the papal bulls partitioning the oceans between the Spanish and the Portuguese (Francis is reported to have asked 'I would like to see the clause in Adam's testament that excludes me from the partition of the world'), sailors from northern France engaged in pillaging Spanish ships in the Caribbean already from 1506 (Toussaint, 1978: 11). And in 1530, it is reported that Francis signed a letter of marque authorising *corsairs* against the Portuguese (Toussaint, 1978: 12). However Protestant they were, many of the captains were not properly licensed. Without licence – or letters of marque, as they would later become known – early exploits were rightly branded piratical by the Spanish.

Yet it was only when the Reformation gained traction in France that privateering developed as a fully-fledged institution of violence. As Mickaël Augeron has noted, for unknown reasons, Calvinist teachings resonated well with sailors. A contemporary mused over the fact in 1584 too, noting that 'almost all sailors of France are Protestants' (Augeron, 1997: 157). This made the task of recruiting sailors an easy one for Huguenot captains. The first letters of marques were not granted until a few months after the Huguenot leader Coligny was made Admiral of France by Henry II in 1553. During wartime against Spain, Huguenot captains would tirelessly attack Spanish ships on their home voyage from the Americas. The end of hostilities between the French and Spanish crown at the Peace of Cateau-Cambrésis in 1559 did not, however, mark the end of this embryonic Protestant hunt for Catholic ships at sea. In fact, many Huguenot captains converted to piracy, still well-protected by their home bases, and hunting Spanish ships in peacetime just as they had done under hostilities. Although the Crown officially complained about this activity, it could not go unnoticed that the Huguenot effort at sea contributed to dramatically weaken France's traditional enemy by attacking the profits from their colonial ventures. 'Unlike the Dutch rebels, Huguenot privateers had operated since the 1560s on the legal edge of French maritime government, acting independently of the crown but formally under the recognised authority of the admiralty of Guyenne, which was held by Navarre' (James, 2004: 21).

Neatly summarised by Augeron, 'Spanish hegemony in the Atlantic was thus contested by a *permanent maritime guerrilla*'

(Augeron, 1997: 161; emphasis added). During peacetime, French captains were granted licences to trade, but these were in fact often hidden orders to pry on Spanish ships. The aim of this was to 'dispute Philip II's mastery of the seas while at the same time avoiding a military confrontation between the two countries' (Augeron, 1997: 161). French Huguenot pirate captains thus operated with total impunity in France, although the consequences of being captured by the Spanish were dire, while French authorities could deny any involvement in the face of the many official Spanish complaints. Yet these Huguenot 'pirates' were not all that piratical, as they were discerning in their prize-taking: they never attacked French ships or allied shipping (Augeron, 1997: 161).

Privateering was to increase in 1562, as Le Havre was returned to the English. The port became a real hotbed of privateers, who attacked Catholic shipping of any nationality in European waters. These privateers seem to have been as much French as English and Dutch, all Protestants united by their common hate of Papists (Augeron, 1997: 163). La Rochelle, the key port of the Huguenots, was 'independent' from 1568–73, and again in revolt a number of times in the 1620s. In order to better control Huguenot privateering as well as to encourage more captains to engage in the activity, letters of marque signed in La Rochelle were introduced. These gave the ongoing quasi-piratical activity an aura of legality, as the letters were signed by a sovereign, the Queen of Navarre Jeanne d'Albret – one of the leaders of the Huguenot party. It should be noted that Coligny seems to have continued to grant more personal notes to certain captains, encouraging them to attack Spanish and Portuguese shipping only and not staying in European waters. Yet, the main objective of Huguenot privateering remained the same: attack the enemies of the reformed religion, defend the Huguenot cause in France, and the 'advancement of the true religion' (Augeron, 1997: 165).

The letter of marque branded the activity as a more legitimate endeavour, and it also gave Huguenot captains a certain guarantee that their prizes would be deemed legitimate as the Admiralty in La Rochelle, at the advice of the Queen, seldom qualified prizes as anything but 'good prizes of enemy and papist goods' (Augeron, 1997: 167). Furthermore, it also gave them protection in any Protestant port, as a condition of the letter of marque was that they did not attack Protestant shipping. As to the Huguenot cause more generally, it ensured that revenue would flow to the Huguenot cause, not least after the 10 per cent of the prize due to the Admiralty was supplemented with 'another Tenth to be employed to the profit of the [Protestant] Cause' (Augeron, 1997: 167). Thus reinforced, the Huguenot privateering effort came to deliver serious blows to the Spanish both in Europe and in the Americas, taking not only bullion from the Americas, but also weaponry destined to the Spanish war effort in the Netherlands. Spain, having to defend herself at sea on several fronts, turned out to be too weak to withstand the Huguenot effort. Close to thirty Catholic prizes were brought to La Rochelle each year in 1570 and 1571, compared with close to 200 for the whole 1536–68 period. As noted above, official Spanish protests seldom lead to the restitution of prizes, and even in cases where the French courts would adjudicate in favour of the Spanish, the Huguenots would categorically refuse to return any prizes (Augeron, 1997: 171).

If letters of marque were not forthcoming from La Rochelle, the Huguenot privateers turned to William of Orange. Huguenot forces also supported William of Orange's campaigns in 1568 and 1572 (van Nierop, 1995: 30), and there had been many plans for a joint venture between Dutch and Huguenot privateers. Furthermore,

> During the third civil war (1568–70), when La Rochelle was at the centre of the Protestant leadership's military initiatives against the crown, their efforts were remarkably successful. The Huguenots were able to put a well-organised fleet to sea ... their forces were in the English Channel and even fighting in the Canaries against the Portuguese. This success cannot be considered in isolation from the near contemporaneous outbreak of the Dutch Revolt with which it was closely involved. From 1568 Louis of Nassau spent four years in France, whilst the Sea Beggars used La Rochelle as a base and received active support from the Huguenots. (James, 2004: 14–15)

As the letters of marque were now issued by William of Orange, so were the sale of prizes in La Rochelle. The French crown had little say in the matter, and Spanish protests were in vain. The end of the third war of religion in France also did not see the end of privateering. Fearing what may happen if hordes of sailors were disbanded, the French crown seems to have preferred that France continue to wage a piratical war on others, rather than on their own citizens (Augeron, 1997: 173). For the Huguenot anti-papist interest aligned with the French anti-Spanish interest, contributing in the long run, as Coligny wrote in 1571, 'the means to render traffic and commerce undertaken by the sea free and safe' (cited in Augeron, 1997: 173).

The formal role played by Henry of Navarre in Huguenot privateering implied all sorts of challenges as he moved closer to the French throne. The privateering of La Rochelle was in a sense caught between the national political ambitions of Henry, the local political ambitions of the Calvinist leaders and the mercantile interests of the merchants of La Rochelle. Securing won positions and reviving trade became more important than supporting privateering. However, 'Navarre continued to issue *congés* [privateering commissions] to captains to pursue prizes from foreign Catholic powers and from anyone not holding a *congé* either from him, from Condé or from his agents in England and the United Provinces' (James, 2004: 21).

Through protecting the Calvinist cause in France, through effective challenges mounted to Spanish shipping in the Americas, and through vast networks of cooperation, and not least due to its successful institutionalisation, Huguenot privateering had become an example to follow for Protestant powers throughout Europe. The model would spread fast, to Dutch and English Protestant parties and beyond.

English privateering

The first years of Elizabeth's reign (1558–1603), when she had to make sense of how to handle the alliance with Spain she had inherited from her Catholic sister Mary, were marked by wariness to engage openly against Spain (Pettegree, 1988; Trim, 1999, 2005). Traditional accounts also highlight the extent to which even a small increase in maritime hostility was sought to be avoided by Elizabeth for fear of escalating the conflict with Spain. Yet, religious sentiment combined with the mercantile drive to extend trade and the predatory drive for plunder of the gentry (Brenner, 1972) made Spain the focus of increased maritime activity. Privateering was the framework through which the drive was channelled (Williams, 1975).

Yet, the privateering of that period was not markedly different from earlier practices, and their reach was rather local. In the 1560s

> most of this marauding was small-scale, not markedly different from petty piracy either in law or in its conduct and results. The ships ... operated chiefly from the Channel ports ... making short trips in home waters to bring in fishing boats, Flemish hoys and such small prizes. The booty was sold along the coasts ... or at recognized plunder marts like Mead Hole in the Isle of Wight ... [M]ost [captains] would be able to produce some sort of commission from a foreign prince. (Andrews, 1966: 16)

It would be inspiration from their Huguenot Protestant brethren which led the English farther to sea (Rodger, 2004). Letters of reprisals were distributed freely to English privateers by the princes of Orange, Condé, and Admiral Coligny, and even Henry of Navarre, leading La Rochelle and the southern ports of England to become 'recruiting and victualling centres as well as plunder marts for Protestant ships which conducted an increasingly indiscriminate war on all "papists"' (Andrews, 1966: 12). This gave an unprecedented boost to English privateering, channelling it against the Spanish, and led to a strong faction of wealthy merchants and shipowners to advocate strongly for a more aggressive English policy against Spain, while pursuing their privateering activities on the Protestant side. With Elizabeth's intervention in France towards the end of 1562 and the deterioration of Anglo-Spanish relations leading to the suspension of trade with Catholic Flanders, English privateers were now granted 'proper' English licences, also against Spanish shipping.

The outbreak of the war in 1585 saw the Iberian merchant interest join the privateering activity, which until then had been characterised by its identification with the Protestant interest. As privateering became a way for merchants to get their hands on produce which was no longer available to them through commercial activity, trade and plunder became intrinsically linked (Shammas, 1978). As Kenneth Andrews puts it, 'Protestantism, patriotism and plunder became virtually synonymous' as the 'petty Channel roving of the earlier years' was turning into the large-scale oceanic ventures of the 1570s and 1580s. By 1585, 'this bellicose alliance of revengeful traders and rapacious gentry gained the adhesion of a powerful body of merchants, and the social force which was to dominate the sea reached its full shape' (Andrews, 1966: 18). The extent was also significant. During the years 1589–91, for instance, Andrews estimates that 'at least 300 voyages – a hundred a year – took place' (1966: 33). Furthermore, reprisal and trade often went hand in hand, and it was not uncommon for a ship to engage in one type of voyage one year, another type the next, or even 'combine trade and plunder in the same venture' (Andrews, 1966: 34). In the context of privateering as a survival strategy, it should also be remembered that the English fleet facing the Armada in 1588 consisted not only of naval vessels, but also a number of privateers.

While the 'international' dimension of the granting of letters of reprisals had allowed the initial predation on Spanish shipping and interests, it had also led to the strengthening of the Isle of Wight as a base for both French, Dutch and English privateers. From there, the first ventures over the Atlantic and the first attempts to establish a privateering base in the Atlantic were launched. The 1570s and early 1580s thus saw an increase in expeditions whose purpose was a mix of the previous anti-Catholic drive, discovery, trade and colonisation. Patronage of privateering expeditions which looked more and more like piracy was no longer concealed by powerful magnates. Privateering had now become the focal activity of different social strata and different economic interests who were moved by a combination of hostility to Spain and Roman Catholicism, the desire for overseas land, exploration, or trade (Hussey, 1929: 287). But as Andrews reminds us, 'the dominant note was predatory' (Andrews, 1966: 18).

Dutch privateering

Privateering from bases in what was to become the Low Countries can be dated at least back to the middle of the thirteenth century, and seems to have originated in much the same way as privateering more generally (Gelderblom, 2005; Lunsford, 2005: 9; Ritsema, 2008: 7). The Dutch Revolt, starting in 1568, plainly changed the scope and meaning of such privateering. In the period covered here, Dutch privateering came in two different forms, somewhat temporally sequenced. The first form of privateering related to the immediate struggle for survival in and around the Dutch provinces, and later to the European theatre of war; the second one came some-what later, and was related to the worldwide efforts to establish trade and fight the Catholic powers. These two forms were logically connected through the overarching aim of weakening the opponents of the Dutch, but they took relatively different forms and were organised and regulated in very different ways. The change from one to the other has been described as a 'change from a *defensive* privateering, practiced by the "watergeuzen" in the early years of the Dutch Revolt to defend their new freedom, to a more *offensive* Dutch privateering to establish a powerful naval force and to encourage a massive colonial expansion' (Kempe, 2009: 394–95; emphasis original).

When William the Silent and others rebelled against Spanish rule in 1567–68 and were driven into exile, they obviously had no navy with which to carry out raids and fight at sea. On the other hand, neither was there a Spanish navy in the north, Philip II having sold his Dutch navy in 1559. Establishing a presence at sea would thus make it possible for the rebels both to dominate one field of military operations and to confuse the enemy and spread his forces on land, by attacking at varying places with short intervals. The solution, for exiles as well as for William, was thus to turn to privateering. Soon a number of Calvinist exiles joined with fellow Protestants in La Rochelle and in English Channel ports (Davis, 1973: 76), cre-ating what amounted to an irregular Protestant armada. As Glete (2000: 153) writes about the Dutch:

> They co-operated with French Huguenot privateers and English privateers with French letters of marque and together they could use English and French ports, as well as Emden in Germany, as bases. During the years 1568 to 1572 this coalition of privateers cruised from the North Sea to the Azores and Canaries and made Western Europe into a war zone where no seafarers, Spanish, Dutch, French or neutral were safe if they were Catholics, loyal to Philip II or carried enemy goods. The attacks were extended to the Caribbean and rumours of a great Protestant attack against the Iberian and Catholic empires circulated.

The Dutch privateers became known as 'Sea Beggars' (*Watergeuzen*), and while some of them indeed engaged shipping on the high seas, many more concentrated mainly on coastal waters and raiding of coastal towns. Lunsford (2005: 4) describes them as 'a ragtag assemblage of Dutch aristocrats, ultra-Calvinists, and riffraff who could be quite indiscriminate as to whom they victimized'.

In addition to the noted French Huguenot letters of marque, William the Silent issued letters, starting in early 1569, as did his brother, Louis of Nassau. While William's letters were accepted, by virtue of him being a prince of Orange, Louis' letters were generally viewed as illegitimate. Even in a friendly country like England, letters from Louis could be declared invalid, although the privateers carrying them were often allowed to leave with their spoils anyhow (Ritsema, 2008: 15–17). While William wanted to create a regular navy out of the Sea Beggars, and forbade the plunder of neutrals, the Beggars needed booty for survival, and alternated between privateering and piracy according to their needs (van Loo, 1997: 175).

In 1572, as a result of negotiations with Spain, Elizabeth I declared that the Sea Beggars would no longer be welcome in England. An immediate result was the first sustained attacks on Dutch land, with the occupation of Brielle and Vlissingen, leading to the subsequent more or less continuous occupation of parts of coastal Holland and Zeeland by 'free' Dutch forces, and steadier forms of income. Securing home bases in the Low Countries and the capacity for buying more ships also in turn meant that the forces gradually became more regularised; Admiralty boards were set up, and a number of former Sea Beggars were made officers, and one even Admiral, of the emerging Dutch navy. Dutch victories in 1573 and 1574 implied that the Dutch and English were masters of the Channel and the North Sea, and that there was no threat from the ocean. From the late 1580s there was furthermore no longer any immediate threat of destruction against the 'heartland'; growing military strength, inland defences and the possibility of inundation gave Holland and Zeeland an island-like quality (Sluiter, 1948: 165–66). Even so, there was no immediate turn to privateering in distant waters.

The lack of overseas engagement was due to the domination of the Dutch in European maritime trade. Apart from an interlude in 1585–88, when Spain was preparing the Armada, Dutch intra-European trade on the Iberian continued largely unabated between 1568 and 1598. Both parties knew that putting an end to this trade would probably cripple the Dutch economically, but that it would also completely cripple the Spanish navy, which, as with all European navies of the time, relied on Baltic timber and naval stores, brought to Spain by Dutch ships. However, two drivers eventually made the Dutch look farther and consider aggressive action. First, some privateers were able to capture Iberian ships on the way back from the colonies. These prizes were enormously profitable, leading to the realisation that it might be easier to pray on such shipping if one sought it at the start of the journey, rather than at the end. Second, there was an inherent risk in relying on the Iberian trade. The trade could be stopped at any time, leading to shortages and dire economic straits, and even if uninterrupted, it made sense to find a way to obtain the goods without middle men. Thus, from 1585 and onwards, when Dutch Iberian trade was first stopped, the Dutch were exploring the East and West Indies commercially (Sluiter, 1942: 29). And most of the merchant ships would make sure to bring a privateering commission, just in case.

While, as we shall soon see below, privateering in colonial waters expanded greatly from 1598, there was little comparable growth in Dutch home waters at the same time. This was to change when the English connection was once again put into play. With the ascension to the throne of Jacob I in 1603 followed peace between England and Spain, and thus an end to English privateering. However, the outfitters and privateers were loath to give up this very profitable business. Thus, from 1604, they turned their attention to the Dutch ports, particularly the ports of Zeeland, where proximity to England and an expatriate English community led to a strong influence on the growth of Zeeland privateering, and 'Of all booty brought into Zeeland in 1605, more than 60 per cent was captured by English privateering ships, vessels which were in some cases manned by Englishmen only' (van Loo, 1997: 187). Partly as a result of a Spanish and Portuguese colonial counteroffensive, the Dutch were to follow suit in 1605 and even more in 1606, when no less than 130 Dutch privateers took to the Atlantic (Sluiter, 1942: 34). The general explosion of privateering put tremendous pressure on Spain, but also caused some problems, namely that a number of privateers were completely indiscriminate in which ships they captured, in practice turning to piracy. The interrelations of different kind of maritime predation can also be illustrated by what happened when the negotiations leading to the truce of 1609 started in 1607. The negotiations implied a stop in privateering, and a number of captains chose to take their talents elsewhere, continuing their careers as pirates operating out of Ireland, or even moving to Barbary (van Loo, 1997: 191), helping set in motion the Atlantic activity of the Barbary corsairs.

Towards a global engagement

After the turn of the century, the commercial climate had changed. This was largely the result of a changing outlook from the side of the English and the Dutch. Both now counted experienced privateering captains and financial backers with a number of successful Atlantic and Eastern ventures under their belt, and as a consequence the range of their ambitions were widening. 1600 saw the first voyage of the English East India Company, a venture largely backed by privateering capital (Chaudhuri, 1965). Henceforth, large-scale London merchants – most of whom had built their fortunes on privateering during the Anglo-Spanish war – became the most important protagonists at sea. London merchants

> did not merely set forth their own vessels. They dominated the whole business of sea-plunder, collaborating with each other, financing amateurs ... working hand-in-glove with the professional seamen, buying up prize cargoes. In all its ramifications and diversity privateering was largely knit together by their ubiquitous activities. ... Thus there developed what might be called a 'privateering interest'. (Andrews, 1966: 120)

While such a description of the privateering world as 'a loose agglomeration of persons connected by their common pursuit' is not unlike what may have developed around most ports or shipping ventures, what set privateering apart was the great influence of those involved: 'They were among the élite of late-Elizabethan and Jacobean merchants' (Andrews, 1966: 120). As Andrews points out, many of the most wealthy and influential merchants of London – who were central in backing the Jacobean state finances – were part of this 'privateering interest': 'every approach points to the conclusion that the riches of these very rich men were built in large measure on the spoil of Spanish and Portuguese shipping' (Andrews, 1966: 121). This wealth, in turn, was what made the Jacobean projection of English power over the seas possible (see

Quinn, 1974); in many respects it not only provided the impetus for sustained English ventures on the sea; it was the condition of possibility of these ventures (Andrews, 1966: 122).

The war against Spain had paved the way for a new form of financial venture, which laid somewhere between trade and predation; a 'parasitic growth'. And although the war against Spain was the longest war since the Hundred Years War, 'it marks the entry of the merchants into the sphere of state finance for the first time on any considerable scale. But what is peculiar to this war is that the merchants did not merely finance the state to wage it, but themselves undertook the waging of its maritime phase for their own profit. The gains they made ... were harvested directly and at no cost of political dependence' (Andrews, 1966: 123).

While the nature of the prizes brought into England came to have a lasting effect on English state finances, it also came to influence the nature and orientation of English trade (Davis, 1973). For as privateering increasingly turned to the Atlantic, not only did it replace the Iberian trade of the pre-war years, but it shifted the emphasis of English imports towards colonial and non-European goods – in spite of England not having any colonies (Lorimer, 1978).

Although the English for a long time failed to establish more than a temporary presence in the West Indies, these nevertheless became one of the main 'hunting grounds' for privateers. This both accompanied and contributed to weakening the Spanish hold on the spoils of Tordesillas. As the Spanish had shown their determination to crush incursions into their domains with the destruction of the French Florida colony in 1565, English efforts proceeded carefully. Before the war broke out, English intentions could not be openly aggressive, and focused mostly on trade. However, even this was done within a language on nationalist economic policies which over time came to acquire a distinct anti-Spanish or even predatory and aggressive tone (Hussey, 1956). In 1577, for instance, Sir Humphrey Gilbert (who was subsequently involved in a number of privateering and colonising ventures) published his *Discourse how her Majesty may annoy the King of Spain* in which he made the case for the capture of Spanish and Portuguese shipping 'either by open hostility, or by some colorable means, as by giving of licence under letters patents to discover and inhabit some strange place'. But these plans did not bear any fruits until more open hostilities with the Spanish. From then onwards, privateering was the complement of nearly every Virginia venture: 'Privateering could cover the expenses of colonisation, and in the circumstances it was the obvious, indeed the only way of maintaining such an enterprise. Since the state would not underwrite it, the Virginia experiment had to pay for itself' (Andrews, 1966: 192). Yet, while the two were propelled by the same logic and had followed the same trajectory, their results were quite different. While English colonial projects kept failing, their prize-hunting was demonstrably profitable. The two thus came to part in 1587, as the focus of the colonisation efforts moved from keeping a permanent garrison to support privateering towards creating a self-sustaining colony of settlers.

Although the early colonial ventures failed initially, they did lead to a more sustained English presence in the Caribbean to the point where the English around 1595 were joining alongside the French in trading in the West Indies. While the Dutch soon outstripped them, the situation in 1595 was tellingly summarised by the treasurer of Santo Domingo:

> corsairs are as numerous and assiduous as though these were ports of their own countries. They lie in wait on all the sailing routes to the Indies ... Coming or going, we always have a corsair in sight. Not a ship coming up on the outside escapes them; nor does any which leaves the harbour get past them. If this continues, either this island will be depopulated or they will compel us to do business with them rather than with Spain ... They make their incursions safely and find persons with whom to barter, for the land is sparsely settled ... Therefore the way to avoid the damage done by these thieves and culprits is to drive the corsairs from these seas. (in Andrews, 1966: 171)

Andrews concludes that 'whatever improvements the Spanish made in their defence of the West Indies, they did not diminish the onslaught of the privateers' (Andrews, 1966: 182). In the early 1600s, most of Hispaniola was 'given over almost entirely to what the Spanish called "rescates" – illicit dealing with the enemy or other unlawful contrabandists. Of the latter many were Dutch and French, who for the most part confined their activity to trade' (Andrews, 1966: 183). But while English privateers also traded with local people, they had less opportunity to do so. Especially compared to the French and Dutch, who came with European supplies for the purpose of trade, English privateers fell short. And although English attempts to emulate French and Dutch trade ventures went well between 1602 and 1603, the end of the war with Spain marked the end of these ventures (see Andrews, 1974).

English privateering, having emerged as an ad hoc tool to help coreligionaries in their continental challenges to Catholic Spain, had by the turn of the century become the main tool of maritime engagement of England. In the process, England had also turned from a polity focused mainly on securing the domestic strengthening of its reformation to a power that challenged Catholic supremacy wherever its ships would carry it. This had happened without any serious ideological engagement. Granted, a number of writers and pamphleteers had engaged themselves in favour of a more sustained and ideological challenge to Spain. However, there had been little response from a queen and her council aware of the limitations of their military power. The importance of the practice of privateering cannot be overstated in the case of England. While promoters of overseas ventures lacked the means to launch large colonial ventures in the 1580s, two decades of privateering war against Catholic shipping had provided these means at the turn of the century. While English seamen lacked experience of the Atlantic in the 1580s, that knowledge, including local knowledge of the West Indies, was common two decades later. While the English merchant fleet was modest before the war, it had experienced a boom over two decades (see Davis, 1962).

Privateering in colonial waters

The main turn towards offensive privateering came when Phillip III issued a trade ban against Dutch commerce in 1598, leading the States-General to systematically start issuing privateering commissions ('t Hart, 2014: 136). The first Dutch forays into colonial waters had come some time before this, with merchants sailing to Brazil from 1587 and onwards, to Guinea in Africa from 1592 to the West Indies from 1594 and to the East Indies from 1595. It has been suggested that the turn to privateering was partly driven by the economic needs of the Dutch navy and state, but the data are inconclusive (Borschberg, 2013: 37–8), and it would seem that private economic interest also factored in, as merchants were suffering immediately from the Spanish ban. The growing competition also implied that a number of captains were returning with little cargo, but with all the more incentive to 'raid and plunder Iberian vessels or their Asian allies as a way of making up for lost time and poor cargo loadings' (Borschberg, 2013: 38).

In the period under discussion here, the most important colonial privateering ground was in Asia (see Borschberg, 2003). This was where most booty could be found, and where Dutch organisation was strongest, with the East India Company (VOC) being founded in 1602. In its heyday, the VOC was the worlds' biggest company, but, as noted by Emmer (2003: 7): 'The commercial success of the VOC has obliterated the fact that, between its foundation in 1602 and the truce with Spain in 1609, the Company was much more an instrument of war than one of commerce.' This was clear already from the charter of the company, which, among other things, gave the VOC a right of defence which came close to an open-ended acceptance of privateering against Spanish and Portuguese shipping.

The move from defensive to offensive privateering, and to a general, unlimited form, did cause debate in the United Provinces, as did a number of spectacular specific cases.[3] From the perspective of International Relations, the writings of Hugo Grotius on the matter are of particular interest. In a long manuscript, known as *De Jure Predae* (On the right of capture), which remained unpublished (and indeed unknown) until the nineteenth century, but from which the 1609 *Mare Librum* was drawn, he discussed extensively when and how states could turn to privateering.[4] As Thomas (2003: 371) summarises the argument:

> De Groot argues that the Portuguese prohibition of trade was in violation of nature, the system of exchange and the fellowship of man. Barring someone from commerce provides a just cause for war. The ocean is a *res extra commercium* and no one has jurisdiction on the ocean. In consequence the VOC was entitled to wage war to exercise its rights, recover its property, losses, expenses, lost profits and to exact punishment for the misdeeds committed against it, its personnel, property and good name. This debt owed by the Portuguese state may be exacted from Portuguese individuals, even innocent merchants. The captured spoils become the property of the author of the war to the amount of the debt.

Crucial to this argument was the attitude that prize-taking was legitimate in just wars, and that wars could be just by relying on defence, recovery of the captured and punishment. Thus, prize-taking became a way of upholding the law, 'in seizing prize or booty, we are attaining through war that which is rightfully ours' (Grotius, cited in Kempe, 2009: 393). Grotius also made the case for general privateering, as opposed to the more traditional redress of individual losses, arguing that there were three just causes for taking prize: '1. compensation for our property unjustly taken by the enemy whether by recovering our own property or an equivalent to the extent of the said property; 2. punishment of offences; 3. indemnity for all losses and expenses of the war' (Kempe, 2009: 393). This was indeed a wide definition, and Grotius even argued that the state could give a general right to plunder (or privateering) to all of its subjects, without issuing commissions. It bears noting that Grotius held that any obstruction of trade legitimated prize-taking, and that the lack of actual Spanish possession (*possessio*) in large parts of the colonies in Asia and America implied that they could not block trade there (van Ittersum, 2007: 61). Although Grotius obviously was aware of the Protestant roots of Dutch privateering, it was important to him to ground the right of prize-taking in natural law and natural rights rather than in religion (van Ittersum, 2007: 84).

In the first twenty years of it existence, the VOC claimed around 200 prizes, worth between 10 and 20 million guilders (Borschberg, 2013: 336, n.2; 't Hart, 2014: 137), and about half of the company's revenue stemmed from privateering (Emmer, 2003: 7). Around the end of our period, the VOC transitioned from privateering to trade, moving 'from a strategy of securing crucial positions in the spice trade to one of developing strong footholds in the main hubs of Asian trade' ('t Hart, 2014: 139–40). This move was possible partly due to superior Dutch organisation and capitalisation, partly due to the massive revenue gathered from privateering. In short, the gains made from privateering had been essential for setting up the economic and military infrastructure which allowed for further expansion in the ensuing periods (Borschberg, 2013: 52).

There is some disagreement as to the overall efficacy of Dutch privateering during this period. Van Loo (1997: 191) argues that Dutch privateering in the Atlantic and in colonial waters was successful enough for it to force Spain into the negotiations leading to the twelve years' truce (van Loo, 1997: 191). Emmer (2003: 12) on the other hand holds that 'Dutch privateering and trade both in Asia and the Atlantic between 1585 and 1609 did not seriously affect Iberian trade and production outside Europe'. However, Emmer builds his observation primarily on raw numbers, finding that Dutch activities at most hindered Iberian growth. This strictly materialist argument ignores first the material point that the very existence of privateering tended to increase transaction costs, by driving up marine insurance rates and by forcing polities to invest in protection schemes. In raw numbers, the Iberian trade might have held up very well, but the maintenance of the status quo was quite obviously much more costly in this period than in previous periods, including enormous defensive expenditure and colonial policies which were detrimental to the colonies themselves (Sluiter, 1948: 195–96; Borschberg, 2013: 40).[5] Second, a strict materialist focus leads to disregard for the psychological impact of privateering. While most ships were not attacked, fear of being attacked was very real and with tangible effects on insurance rates, state policy and marine activity in general. The impact of Dutch privateering extended way beyond the material. Iberian merchants and polities which were used to unimpeded growth suddenly had to fight to even retain the status quo. This again ties in with the third factor missing from the materialist argument, the ideological challenge against Iberian legitimacy. Tordesillas had divided the world between the Iberian polities, and Spain and Portugal accordingly labelled anyone challenging this state of affairs as pirates. However, such labelling proved ineffective. Dutch (and English) privateering undercut the claim to exclusive rights, and thus also the Catholic world order. Seen in this perspective, Dutch (and English) privateering proved wildly successful, allowing not only for survival of the polity and the accumulation of capital at home, but for a thorough change in the conceptualisation of the globe.

In the first period discussed here, Dutch privateering was a weapon of the weak, a virtue of necessity and a last resort. Whatever existed of regulations and control was completely subordinated to the necessity of winning, but only until bases had been established. Even before a centre of gravity was established, William the Silent tried to discipline his privateers. In later phases, when the threat was less acute, privateering was much better regulated, but also more strategically put to use in the global war against the Iberian powers. Rather than being the weapon of the weak, privateering was the weapon of the upstarts – states as well as individuals. Through the initiative of thrifty merchants and the VOC, Dutch privateering helped bring about the golden age of Dutch imperialism. If the first phase was about privateers going to sea to save the homeland, the second phase was about privateers going overseas to claim new land.

Conclusion

As we have demonstrated, Protestant privateering was not only a distinct phenomenon; it decisively shaped the trajectory of global politics. The connections between the different Protestant privateering communities were close in the decades between the 1560s and 1607, including joint ventures, a broad understanding of who could grant whom privateering commissions and a strong loyalty based on the existence of a common enemy. There is little doubt that religion and the religious wars provided the backdrop for the intensification of privateering, and as such came to inform the practice throughout our period in terms of defining friend and foe on water. To be sure, privateering expeditions were mounted on the basis of expectations of financial gains. However, many of them came back with accounts of how Catholic churches had been burnt, friars hanged, and crucifixes broken. The religious wars were transposed to the largely apolitical space of the sea, providing in effect not only the motivation to go to sea, but the structure of overseas expansion.

In general, however, motivations were mixed, and the different instances of privateering discussed above illustrate vividly how privateering was a complex phenomenon where public and private interests were intertwined and intermingled. Furthermore, any notion of long-term planned developments is put to lie by the empirical account. Privateering arose as a virtue of necessity for survival for the Huguenots and in the early Dutch Revolt, as opportunistic challenge against the hegemons of the sea by the English

Sea Dogs and as support for, and alternative to, trade in the colonies, as demonstrated in all of the cases. Once developed, it proved to be much more effective than the initial quasi-pirates must have imagined, providing the Protestant cause with the means to wage a constant war upon the papists – even in peacetime – or, as Augeron called it, a 'permanent maritime guerrilla' upon the Catholics (Augeron, 1997: 161). This had effects far beyond the weakening of the Spanish colonial presence. It reduced Spanish revenue – at times dramatically – making it more difficult for Spain to wage its war against the Dutch and forcing them into a defensive posture in their colonies. Taking into account the sea, and especially the extent to which modes of violence at sea and on land were far from correlated, allows us to address an important cause of the strengthening of Protestantism not only in Europe, but also beyond European shores.

Traditionally, privateering has been dismissed as being without real consequence for larger questions of power distribution and overseas exploration and colonial ventures. Such a conclusion is problematic, however, as it overlooks all the prizes that were taken on a routine basis by a host of ships engaged in privateering, as well as the extensive system of convoys and protection set up by the Iberian powers to protect their shipping. Assessing the importance of privateering in terms of their prizes is also difficult, as a varying percentage of the value of the inventory had to be paid to the Admiralty; cargoes were often disposed of before they had been inventoried. Furthermore, assertions about the importance of privateering based on the fact that no important captures of bullion were ever made overlook the fact that privateers seldom met or challenged warships. Privateering was not about inflicting massive damage to enemy ships, but about capturing them in close fighting. As such, the vast 'majority of prizes were of the middling and lesser sorts of merchantmen, easy prey to any privateer' (Andrews, 1966: 39). Privateering mattered primarily not through spectacular captures, but through the steady 'death by a thousand cuts' and the psychological, ideological and juridical challenges against the Iberian hegemony.

Protestant privateering marks a turning point in the history of privateering and is of particular importance due to its transformative quality. It shows how a medieval institution was tweaked in order to serve the interests of a new order, an order defined by its radical rupture with the religious unity of the Middle Ages. The first major consequence of Protestant privateering was thus the very survival of Protestantism in north-western Europe. Even though the Huguenots were eventually losers, England and the Netherlands remained, ensuring that Protestantism was to thrive even outside the relative backwaters of Scandinavia and Northern Germany. The second major consequence flowed from the first. While privateering had previously been little more than an incidental concomitant of war, it now became the main driver of overseas expansion, which came to set its mark not only on the means through which European colonial ventures were financed and undertaken, but on the global settlement of trading posts and colonies. In short, it came to define the main structures of the colonial world. Later forms of privateering would again work within the normative structures of oceanic governance, but these were new structures established through Protestant privateering. Finally, it is worth noting how the end of Protestant privateering, through migration, in turn led to an uptick in the privateering activities of the 'Barbary corsairs', allowing them for the first time to venture into the Atlantic.

Our discussion in this chapter also demonstrates the necessity of taking the sea seriously in our analyses of international relations. One could tell the story of privateering in terms of the tropes of taming or traversing, with privateering as a vehicle for mastery at sea or as a way of getting across the sea. It was both, but it was also much more. Privateering helped change the very conceptualisation of the sea and its relation to land, making it impossible to think of the sea as something which could be fully controlled or merely a space to traverse. This reconceptualisation of the sea would last until the establishment of British naval hegemony, which, perhaps not coincidentally, coincided with the gradual demise of privateering.

Notes

1 Earlier drafts of this chapter were presented at the ISA's 55th Annual Convention in Toronto, March 2014 and at EWIS Krakow 2019. Thanks are due to our discussants, Torbjørn Knutsen, Richard Mansbach, Julia Costa López and Xavier Guillaume, as well as to our fellow panellists.

2 The chapter draws on a diverse set of secondary sources. To the extent that privateering has been a scholarly topic, it has more often than not been discussed in nationalist historiographies, with very different emphasis in France, England and the Netherlands. Our attempt to read Protestant privateering as a cross-polity phenomenon thus cuts against the grain of much existing literature (but see Rodger, 2004). In International Relations, Phillips (2010) has argued along somewhat parallel lines for the trans-polity character of sectarian conflict. It should be added that no general account of privateering as a historical or legal phenomenon exists.

3 Of particular importance was the capture of the Portuguese carrack *Santa Catharina* in the winter of 1603. The ship, containing enormous riches, was taken by a captain acting without a permission to initiate action from either the state or the VOC, but it was nevertheless declared as a good prize in the autumn of 1604.

4 Grotius was commissioned to write the manuscript as a result of the *Santa Catharina* case, but upon its completion the negotiations which led to the truce of 1609 were already underway, and it was deemed unwise to publish it.

5 From an overall Dutch perspective, it could also be argued that Emmer misses the point. One crucial effect of colonial privateering was that the Iberian powers had less resources to put into the European theatre of war (Borschberg, 2013: 40).

References

Andrews, K. R. (1966). *Elizabethan Privateering: English Privateering During the Spanish War 1585–1603*, Cambridge: Cambridge University Press.

Andrews, K. R. (1974). 'Caribbean rivalry and the Anglo-Spanish peace of 1604', *History* 59: 1–17.

Augeron, M. (1997). 'Coligny et les Espagnols à travers la course (c. 1560–1572): unde politique maritime au service de la Cause protestante', in M. Acerra and G. Martinière, eds, *Coligny, les protestants et la mer*, Paris: Presses de l'Université de Paris-Sorbonne, 155–76.

Borschberg, P. (2003). 'Portuguese, Spanish and Dutch plans to construct a fort in the Straits of Singapore, ca. 1584–1625', *Archipel* 65: 55–88.

Borschberg, P. (2013). 'From self-defense to an instrument of war: Dutch privateering around the Malay Peninsula in the early seventeenth century', *Journal of Early Modern History* 17: 32–52.

Brenner, R. (1972). 'The social basis of English commercial expansion, 1550–1650', *Journal of Economic History* 32: 361–84.

de Carvalho, B. (2014). 'The confessional state in international politics: Tudor England, religion, and the eclipse of dynasticism', *Diplomacy and Statecraft* 25(3): 407–31.

Chaudhuri, K. N. (1965). *The East India Company: A Study of and Early Joint-Stock Company, 1600–1640*, London: Frank Cass and Co.

Davis, R. (1962). *The Rise of the English Shipping Industry*, London: Macmillan.

Davis, R. (1973). *English Overseas Trade, 1500–1700*, London: Palgrave Macmillan.

Emmer, P. C. (2003). 'The first global war: the Dutch versus Iberia in Asia, Africa and the New World, 1590–1609', *e-JPH* 1: 1.

Gelderblom, O. (2005). 'Violence and growth: the protection of long-distance trade in the Low Countries, 1250–1650', Working Paper, Social and Economic History Faculty, University of Utrecht.

Glete, J. (2000). *Warfare at Sea, 1500–1650: Maritime Conflicts and the Transformation of Europe*, London: Routledge.

't Hart, M. (2014). *The Dutch Wars of Independence: Warfare and Commerce in the Netherlands 1570–1680*, London: Routledge.

Hussey, R. D. (1929). 'Spanish reaction to foreign aggression in the Caribbean to about 1680', *The Hispanic American Historical Review* 9(3): 286–302.

Hussey, R. D. (1956). 'America in European diplomacy, 1597–1604', *Revista de Historia de América* 41(June): 1–30.

James, A. (2004). *The Navy and Government in Early Modern France, 1572–1661*, Woodbridge: Boydell and Brewer.

Kempe, M. (2009). 'Beyond the law: the image of piracy in the legal writings of Hugo Grotius', in H. W. Blom, ed., *Property, Piracy and Punishment: Hugo Grotius on War and Booty in 'De iure praedae'*, Leiden: Brill, 379–95.

Leira, H., and B. de Carvalho (2011). 'Privateers of the North Sea: at world's end – French privateers in Norwegian waters', in A. Colás and B. Mabee, eds, *Mercenaries, Pirates, Bandits and Empires. Private Violence in Historical Context*, New York: Columbia University Press, 55–82.

Leira, H., and I. B. Neumann (2008). 'Consular representation in an emerging state: the case of Norway', *The Hague Journal of Diplomacy* 3(1): 1–19.

Lorimer, J. (1978) 'The English contraband tobacco trade in Trinidad and Guiana, 1590–1617', in K. R. Andrews, N. P. Canny and P. E. H. Hair, eds, *The Westward Enterprise*, Liverpool: Liverpool University Press, 124–50.

Löwenheim, O. (2007). *Predators and Parasites. Persistent Agents of Transnational Harm and Great Power Authority*, Ann Arbor, MI: The University of Michigan Press.

Lunsford, V. W. (2005). *Piracy and Privateering in the Golden Age Netherlands*, Houndsmills: Palgrave Macmillan.

Pettegree, A. (1988). 'Elizabethan foreign policy', *The Historical Journal* 31(4): 965–72.

Quinn, D. B. (1974). 'James I and the beginnings of empire in America', *The Journal of Imperial and Commonwealth History* 2(2): 135–52.

Ritsema, A. (2008). *Pirates and Privateers from the Low Countries c.1500–c.1810*, Deventer: Ritsema.

Rodger, N. A. M. (2004). 'Queen Elizabeth and the myth of sea-power in English history', *Transactions of the Royal Historical Society* 6(14): 153–74.

Shammas, C. (1978). 'English commercial development and American colonization, 1560–1620', in K. R. Andrews, N. P. Canny and P. E. H. Hair, eds, *The Westward Enterprise*, Liverpool: Liverpool University Press, 151–74.

Sluiter, E. (1942). 'Dutch maritime power and the colonial status quo, 1585–1641', *Pacific Historical Review* 11(1): 29–41.

Sluiter, E. (1948). 'Dutch-Spanish rivalry in the Caribbean Area 1594–1609', *The Hispanic American Historical Review* 28(2): 165–96.

Steinberg, P. E. (2001). *The Social Construction of the Ocean*, Cambridge: Cambridge University Press.

Sutherland, N. M. (1992). 'The origins of the Thirty Years War and the structure of European politics', *The English Historical Review* 107: 587–625.

Thomas, P. (2003). 'Piracy, privateering, and the United States of the Netherlands', *Revue Internationale des droits de l'Antiquité* 50: 361–82.

Thomson, J. E. (1994). *Mercenaries, Pirates, and Sovereigns. State-building and Extraterritorial Violence in Early Modern Europe*, Princeton, NJ: Princeton University Press.

Toussaint, A. (1978). *Histoire des Corsaires*. Paris: Presses Universitaires de France.

Trim, D. J. B. (2005). 'Seeking a Protestant alliance and liberty of conscience on the continent, 1558–1585', in S. Doran and G. Richardson, eds, *Tudor England and Its Neighbours*, London: Palgrave Macmillan, 139–77.

Trim, D. J. B. (1999). 'The "secret war" of Elizabeth I: England and the Huguenots during the early wars of religion, 1562–77', *Proceedings of the Huguenot Society of Great Britain and Ireland* 27(2): 189–99.

van Ittersum, M. J. (2007). 'Mare Liberum in the West Indies: Hugo Grotius and the case of the swimming lion, a Dutch pirate in the Caribbean at the turn of the seventeenth century', *Itinerario* 31(3): 59–94.

van Loo, I. (1997). 'For freedom and fortune: the rise of Dutch privateering in the first half of the Dutch Revolt, 1568–1609', in M. van der Hoven, ed., *Exercise of Arms: Warfare in the Netherlands, 1568–1648*, Brill, 173–95.

van Nierop, H. (1995). 'Similar problems, different outcomes: the Revolt of the Netherlands and the Wars of Religion in France', in K. Davids and J. Lucassen, eds, *A Miracle Mirrored: The Dutch Republic in European Perspective*, Cambridge: Cambridge University Press, 26–56.

Williams, N. (1975). *The Sea Dogs: Privateers, Plunder and Piracy in the Elizabethan Age*, London: Weidenfeld and Nicolson.

6

A sea of connectivity and entanglement: modern mobilities and ancient thalassocracies in the Mediterranean Sea

Andonea Jon Dickson

The Mediterranean Sea is increasingly defined by the interactions that take place between migrants and the actors working within this sea to assist and contain human mobility. Contemporary scholarship on borders, migration, and security studies have examined the changing geographies of policing at sea (Basaran, 2010; Carrera and Den Hertog, 2015), the influence of border policies on a re-design of humanitarianism (Hyndman and Mountz, 2008; Williams, 2015; Cusumano and Pattison, 2018), and the risk migrants are placed in at sea (Heller and Pezzani, 2012; Stierl, 2016). Absent from these discussions, however, is an analysis of the way the sea is a living geography in its own right and how seafaring is tied in with political ordering. Studying the sea as a geography made through entanglement and connectivity requires an inversion of perspective; instead of a landlocked imagination that situates the sea as a liminal geography, this chapter calls for the sea to be located as a central vantage point. Through this, the sea emerges not just as a distance to be traversed on the way to more terrestrial geographies, but rather as a dense political geography made through various forms of at-sea action and interaction. The chapter considers how the sea is evolving through becoming a site of mobility control, exploring the methods of at-sea connectivity that emerge as a result of migration and its regulation and how power is rendered through these practices.

While the maritime geography has long been a space of political activity and trade, we remain without a language to consider the changing patterns of seafaring that define the maritime realm. The law of the sea is often invoked in considering the control of human mobility at sea and provides grounds for disputes of migration governance. Yet, as this chapter will demonstrate, the law of the sea does not inform us of the way mobility regulation is shifting how this sea is being enacted, leaving us in need of a language to engage with the transforming dynamics within this geography. Maritime security equally does not provide detail on changing connectivities at sea. In order to build up a language that considers how power is rendered through the sea, the chapter turns to the power over the sea that the Greeks held during antiquity, also known as the Greek thalassocracy. While many periods of time, and many world views other than European, could have been used to demonstrate the significance of at-sea connectivity, the Greeks had a distinct view of the Mediterranean, situating the sea as central to the ordering of politics and society. Borrowing on this imaginary of the sea from antiquity, we begin to develop a language that brings the sea to the fore, illustrating it to be a space made through the practices which transpire within it. This dissolves the distinction between land and sea by emphasising the maritime geography to be the site of political activity, where power is rendered through various at-sea practices.

This chapter begins by examining contemporary strategies of mobility control in the Mediterranean, focusing specifically on the practices of Eunavfor Med. Following this, the chapter articulates the need to consider this geography not simply as a space traversed, but rather as a geography defined by the action and interaction within it. From here, the chapter turns to antiquity to build up a language on this sea, focusing on the centrality of the sea and the micro actions that produced this space for the ancient Greeks. From navigational strategies to complex nodal systems, the Greek thalassocracy exposes a way of perceiving the sea through a networked connectivity that is both diffuse as well as feeds into a macro ordering of this geography. Through considering the way this sea worked for the Greeks, we begin to consider the significance of the at-sea practices which continually make and re-make this political geography.

Mobility control in the Mediterranean

It is important to understand how this sea has developed through becoming a site of migration management and the correlating practices that are redefining this geography. The sea is a complex political space, constantly undergoing a process of being made. While we know that the materiality of maritime spaces are 'unmistakably undergoing continual reformation' (Steinberg and Peters, 2015: 248), so too are the policies and practices within these geographies. This is most aptly demonstrated by practices of migration regulation. While the law of the sea is often used as a framework through which to consider at-sea practices, this section demonstrates how law does not illustrate the dynamic reformation of this sea, leaving us in need of a language to consider the changing forms of connectivity informed by strategies of mobility control.

Mobility control in the Mediterranean is presently defined by several major programmes, although these programmes are in a constant condition of change. At the end of 2019, the Council of the European Union cited there to be four EU operations in the Mediterranean focused specifically on securing European borders, disrupting smuggling networks, and rescuing individuals in a condition of risk (Council of the European Union, 2019). These four programmes include three Frontex operations: Indalo, Themis (which replaced Triton), and Poseidon; as well as the European Naval Force Mediterranean (Eunavfor Med), also termed Operation Sophia. Eunavfor Med, explained in more detail below, was a short-term programme which began in mid-2015 and was focused on dismantling smuggling networks at sea. While it was due to terminate at the end of December 2018, the programme was extended to the end of March 2020, yet during the final year of operation it functioned only with an aerial and surveillance capacity, with its naval assets suspended. In addition to Eunavfor Med, there is the interstate military alliance NATO which, under Operation Sea Guardian, has ships and planes patrolling the Aegean and the central Mediterranean, providing 'logistical support' to EU border agencies (NATO, 2016). Each of these programmes regulate human mobility, whether it be in capacity building, surveillance, intercepting vessels, repatriating vessels, transiting individuals to

European shores, or penalising smugglers. These programmes exist in correlation with the military and border control capacities of the Member States that comprise the littoral geographies of the Mediterranean. There is also a significant number of humanitarian operations that have, over recent years, contributed to the complexity of connectivity at sea. While not all are still presently active, these humanitarian operations have included: Migrant Offshore Aid Station (MOAS), Sea-Watch, Médecins Sans Frontières (MSF), as well as Jugend Rettet, Proactiva Open Arms, Sea-Eye, Save the Children, and SOS Mediterranee, among others. While the focus of this chapter is on the policing strategies in this sea, the humanitarian agencies operating agendas of rescue are not outside this discussion of the shifting nature of connectivity at sea.

Although legal frameworks do indicate strategies of policing the Mediterranean, they are limited in the context of explaining the evolving and discursive nature of power in this geography. On the one hand, the law of the sea suggests territorial delineations at sea which lose pertinence in strategies of mobility control. On the other hand, the law of the sea is often addressed in a broad way, especially in relation to security: 'A hallmark of the law of the sea has been the preference to treat security concerns implicitly rather than explicitly' (Klein, 2015: 597). In fact, security in maritime law is often discussed in a 'context-specific' way or in a way which allows states a capacity to negotiate (Klein, 2011: 11). As such, the applicability and inapplicability of maritime law within agendas of maritime mobility control is somewhat open for interpretation. Through considering the effects that Eunavfor Med has had on this sea, as well as the changing practices around rescue, it becomes apparent that the law of the sea cannot convey the messy and entangled way the Mediterranean is being enacted in migration management.

A central understanding of maritime geographies and the operations within them is the delineation between high seas and territorial seas. These principles are grounded in early postulations of laws of the sea. In 1609, Hugo Grotius' 'The Free Seas' or *Mare Liberum* argued for a freedom of navigation at sea, as well as a coastal 'maritime belt' over which a state would exercise authority. These principles of high seas and territorial waters were sustained and are reflected in the United Nations Convention on the Law of the Sea (UNCLOS) 1982. A state now has a right to exert full territorial authority over 12nm of ocean from the baseline of the state, inclusive of the air space and subsoil. States are also entitled to claim a potential 12nm contiguous zone, within which a state has a right to prevent and punish 'infringement of its customs, fiscal, immigration or sanitary laws and regulations within its territory or territorial sea' (Art. 33). A state also has a right to an exclusive economic zone (EEZ), which stretches 200nm from the baseline of a state. Within the EEZ there is a more restricted sovereign authority, largely relating to 'exploring and exploiting, conserving and managing the natural resources, whether living or non-living' (Art. 56 1.A). The high seas remain 'free', so long as they are used for 'peaceful purposes' (Art. 88). These stipulations stitch together to demonstrate the fabric of sovereign geographies at sea and the type of action permitted within these spaces.

In a way distinct to other maritime geographies, the Mediterranean has a variegated jurisdictional geography. There is a curtailment of some of these jurisdictional rights within this semi-closed sea. Not all states have claimed a 12nm territorial zone, with Greece and the Aegean coast of Turkey for example only claiming 6nm, while many states have not claimed a contiguous zone, or EEZ.[1] The effect of this is 'a patchwork of jurisdictional rights and duties of states and the enduring presence of high seas areas that are within 200 nautical miles (nm) of coasts' (Papanicolopulu, 2015: 607). This variance in claimed maritime zones sets the Mediterranean 'apart from all other seas' (Papanicolopulu, 2015: 607). With many of these Mediterranean states restricting their jurisdictional geographies at sea, the Mediterranean has high seas that are more proximate to coastal territories than in other oceans and seas around the world (Cacaud, 2005). This allows for a greater area at sea where there is liberty in resource extraction and, indeed, in movement.

Within the operations of Eunavfor Med, these geographies of freedom of movement lose pertinence. Eunavfor Med commenced operation in June 2015, operating with the mandate of disrupting networks of smuggling in the central Mediterranean. It was a programme comprised of four phases. Phase one consisted of surveillance, monitoring, and information gathering on migration networks and smuggling. Phase two was comprised of two parts: (1) the conducting of 'boarding, search, seizure and diversion on the high seas of vessels', and (2) the same actions executed in the 'territorial and internal waters' of Libya. Phase three enabled the operation to 'take all necessary measures against a vessel' including the destruction of a vessel (European Council, 2015). Lastly, phase four refers to the withdrawal of the operation.

In analysing phase two, the interruption of vessels on the high seas and in the territorial waters of Libya, there emerges an interesting unfastening of the territorial limitations of a European policing agenda through an expanded right of interruption at sea. While there are provisions for a state 'to board a vessel for immigration matters or to suppress a criminal activity' within its own territorial or internal waters, on the high seas it is more ambiguous as carrying migrants on the high seas is not a criminal act in itself (Papastavridis, 2008: 152, 163). UNCLOS does not specifically justify a right to interception; however, there are conditions which highlight a right to the visitation of vessels (Papastavridis, 2008). As stipulated by Article 110(1) of UNCLOS, the right to visitation on the high seas is premised upon the legal right to interrupt in order to prevent '(a) piracy (b) slave trading (c) unauthorised broadcasting (d) absence of nationality of the ship, or (e) though flying a foreign flag or refusing to show its flag, the ship is in reality of the same nationality as the warship'. Thus, within Article 110, there is no clear stipulation regarding the transportation of migrants or even the trafficking of persons (Papastavridis, 2016: 61). However, there is potential for modern migration and trafficking to be interpreted through a lens of coercion and slavery, the debate of which is beyond the ambit of this chapter (see e.g. Papastavridis, 2008; Bevilacqua, 2017). Equally, unregistered vessels transporting migrants are often navigating the Mediterranean without a flag state, which equally provides grounds for interference.

The only multilateral treaty that definitively refers to the interdiction on the high seas of migrants is the Protocol Against Smuggling of Migrants Protocol by Land, Sea and Air (2000). The Smuggling of Migrants Protocol allows for interdictions on the high seas providing smuggling is a concern and permission from the flag state of the vessel has been granted (Papastavridis, 2016: 62). Article 4 of the Migrant Smuggling Protocol states that offences must be 'transnational in nature and involve an organized criminal group' which, as Klein (2011: 125 footnote) indicates, omits the instance of a 'one-off attempt at migrant smuggling, or independent operators'. Furthermore, coercive actions such as redirecting migrant journeys at sea remain unclear. The Protocol does not indicate what specific action is permitted if suspicions are confirmed,[2] but stipulates that further action must correlate with UNCLOS. While UNCLOS allows for visitation, 'it does not explicitly allow the exercise of additional coercive powers, but at the same time it does not prohibit them' (Bevilacqua, 2017: 173). Hence, the action taken with a migrant vessel subject to suspicion of smuggling is ambiguous. This further emphasises Klein's earlier argument that principles of security within maritime law are largely a point of negotiation.

Interdictions at sea become more problematic in discussion of the operations within the territorial waters of Libya, as stipulated by the second part of phase two of Operation Eunavfor Med. The regulating of mobility within the territorial limits of another state has the potential to be interpreted as interfering with 'the fundamental principle of territorial sovereignty', which is not only a condition of international law but is stipulated by UNCLOS (Bevilacqua, 2017: 176). Resolution 2240 by the UN Security Council in May 2015 deemed action within Libyan territorial waters justified on the grounds of the instability of the Libyan state and the seriousness of the condition of smuggling in the Mediterranean (Bevilacqua, 2017: 175). Despite this, the 'political and legal conditions' for the operation of Eunavfor Med in Libyan territorial waters did not transpire, and as such, direct action within the territorial waters of this state has been limited (European External Action Service, 2018: para 88).

While the deployment of Eunavfor Med to Libyan territorial waters has been restricted, training programmes led by Eunavfor Med have had a similar effect of stitching a European agenda to the coastal shores of the southern Mediterranean. In 2016, the Political and Security Committee of the EU concluded that Eunavfor Med should begin tasks relating to ensuring the UN arms embargo and training the Libyan Coast Guard (LCG) (Council of the European Union, 2016). As such, in September 2016, Eunavfor Med launched the additional task of training the LCG. Succeeding this, the Malta Declaration in early 2017 also emphasised that 'priority will be given to … training, equipment and support to the Libyan national Coast Guard and other relevant agencies. Complementary EU training programmes must be rapidly stepped up, both in intensity and numbers, starting with those already undertaken by Operation Sophia [Eunavfor Med]' (Council of the European Union, 2017). In fact, the efforts to bolster migration control in Libyan waters have been multifaceted. The Italian government has also directly contributed to this project; 'From the beginning of 2017 onwards, the Italian government backed by the EU has increasingly cooperated with Libyan authorities to block departures in exchange for financial and logistical support' (Villa et al., 2018).

The training and support of the LCG has become an important part of EU policy relating to the control of migration from Libyan shores. In January 2018, UNHCR evidenced the influence of the European-informed agenda in Libyan waters through the increases in migrants who were disembarked in Libya on behalf of the LCG. In this sense, rescue on behalf of the LCG functions as a form of capture or outsourced push-back (Cuttitta, 2018). In November 2017, 1,214 migrants were disembarked at Libyan ports by the LCG; in December 2017, 1,157 migrants were disembarked in Libya by the LCG; and, in January 2018, 1,430 migrants were disembarked (UNHCR, 2018). Through increasingly containing migrants in Libya, the practices of the LCG demonstrate how a European logic of migration containment is fastened to the littoral geographies of the southern Mediterranean in a way not easily explained through a framework of maritime law. It should be noted that the Libyan government has frequently opposed European policies of migration regulation. Moreover, during certain periods of unrest and civil war in Libya, the LCG stopped operating at sea. These practices are thus not consistent, but rather evidence a sea of entanglements, through which migrants are increasingly contained and restricted.

A sea of connectivity

The Mediterranean Sea is a complex political geography. In addition to the private and commercial vessels that straite this sea, there are multiple programmes operating in this geography to control human migration. These programmes, along with EU policies on mobility regulation at sea, change conditions of interaction in this maritime geography, demonstrated above through a proliferation in interceptions. These practices do not adhere to the limitations of territorial waters, nor do they reflect sanctioned practices of interception at sea. As such, these points of interactivity function as nodes which, while being ephemeral in nature, nonetheless demonstrate entanglements and connectivities that define this sea; they elucidate methods of seafaring that are increasingly informed by European agendas of mobility control. Looking at these nodes of connectivity at sea thus reveals a power that is rendered across this liquid geography.

Through situating the sea as a central vantage point, taking seriously the forms of seafaring that influence this geography, this chapter proposes that we move past conceptualisations of political and politicising acts as being intrinsically grounded in order to understand how these very same processes shape maritime geographies. Focusing on these at-sea practices as illustrating political power ties in with broader critiques of the capacity of International Relations (IR) to critically challenge and move beyond the concept of territory (Agnew, 1994; Shah, 2012). As Shah (2012: 11) writes, 'the juridical conception of the "bordered territory" is reduced to a brute, physical landscape, which effaces but reinforces the normative assumption that territory is the fixed and eternal site of politics'. The argument here is not that the maritime geography is the same as territory; as Steinberg and Peters (2015: 248) articulate, there is a 'material and phenomenological distinctiveness' to the sea. Rather, it is to emphasise that, through taking the sea seriously, we are able consider how power and order emerge through seafaring strategies and at-sea practices. Although the maritime geography is a space defined by movement, it is not simply a geography to be traversed; instead, these mobile actions and interactions reveal shifting dynamics within this sea. While a number of scholars have interrogated social and political constructs such as sovereignty (e.g. Biersteker and Weber, 1996), we need a similar language in IR that addresses the connectivities and entanglements that *make* maritime geographies.

The control of human mobility at sea could be interpreted though nascent discussions on the securitisation of maritime geographies (e.g. Ryan, 2015; Bueger and Edmunds, 2017). Maritime security studies draw on previous concepts relating to sea power, marine safety, resilience and the blue economy (Bueger, 2015: 160). Within this field, attention is commonly dedicated to what constitutes threats at sea, rather than what maritime security specifically might mean (Klein, 2011: 9). Bueger and Edmunds (2017: 1299) have emphasised the recent reconfiguration of the seas through security agendas and the correlative emergence of an 'interlinked security complex' in maritime geographies. The effect of this includes expanded geographies of policing that are more interstate, a diversification of actors, and the realisation that insecurity at sea does not always begin at sea but is often funded and facilitated on territorial geographies (Bueger and Edmunds, 2017). While this opens up IR to a conversation of terraqueous systems of governing, this literature does not provide a way to engage with the dynamic and changing practices at sea. There is something greater than the multiplication of actors and the spectral nature of jurisdictional limitations transpiring through regulating mobility at sea. The European objective of containing human migration at sea is influencing practices and customs within the Mediterranean, reflecting a making and re-making of this geography that is more intricate than what security discourses illuminate. As such, neither security discourses nor legal frameworks capture the connectivities and entanglements which formulate this space as lived and living.

In order to understand the present nodal systems and navigational strategies of the sea as contributing to a transformative geography, I argue that we can turn to studies of how this geography functioned during antiquity for, as Horden and Purcell (2000: 133) write, 'The study of maritime connectivity is one subject in which the Mediterranean historian has most to learn from prehistory.' Turning to antiquity allows for a conceptualisation of this sea as a living geography, possessing a condition of always being made, as well as facilitating an understanding of more enduring logics of a governing of this space that is achieved through strategies of at-sea connectivity. The methods of seafaring during antiquity demonstrate how movement at sea and the connectivity this proliferates can engender a political power and a way of governing *through* the sea.

Navigating antiquity

While studying the Mediterranean through at-sea connectivities is a nascent way to interpret this geography, it is not a new way of operating in this geography. As Horden and Purcell (2000: 21) argue, the Mediterranean has largely been conceptualised as a geography 'ineluctably divided', rather than functioning as a geography deeply defined by connection. Archaeologists and historians of the Bronze Age principally consider this space and the connections within it as a whole, and in so doing, expose the long history of the networks that have formulated this seascape. This is especially pertinent to the Greek thalassocracy. A definitional understanding of the term thalassocracy would be 'sea power', deriving from the Greek *thalassokratia*. Yet, more specific to the social, cultural, and political history of the term, thalassocracy indicates a notion of power that is realised both *on* and *through* the seas. It is from this period that we can develop a conceptual framework through which to consider the political significance of the maritime and the various modes of enacting this space. Through this, we can generate ways to interpret a thalassic authority realised through connectivity at sea.[3] While subsequent thalassocracies, such as that of the Roman or the Arabic Empires, could be used as an analogy to the contemporary politicisation of the sea, these thalassocracies treated the sea as a space between land, a space that affords connectivity, rather than the Grecian way of interpreting the sea a space *of* connectivity (Harris and Harris, 2005; Malkin, 2011). In demonstrating how the sea can be considered as densely political with a power rendered through connectivity, there is thus a pertinence to considering how the Greeks used the sea.

During its height, the Greek thalassocracy extended throughout the littoral spaces of the Mediterranean. As Malkin (2011: 19) explains, from the close of the second millennium to 400 BCE, Greek communities 'dotted along the coasts of mainland Greece, the Aegean, Asia Minor, the Propontis, the Black Sea, Italy, Sicily, France, Spain, and north Africa'. The Mediterranean Sea had effectively, by the fifth century BCE, become 'a single hinterland' (Horden and Purcell, 2000: 134). Whether it be between Naukratis in Egypt and the communities of Rhodes, or the Greeks of Phocaea and their settlement in Agde France, there was a well-established connectivity between these otherwise deeply distant locales. Athenian policies linked domestic and overseas territories, giving rise to complex networks of travel: 'there were places from which to obtain supplies and, just as importantly, places that guarded the longer routes leading to sources of supply' (Abulafia, 2011: 139). The lack of contiguity between these geographies was not simply countered by maritime transport, it was in fact the sea that produced a connectivity; this thalassocracy functioned 'with settlements sprinkled on distanced, disconnected shores, and with the "empty" sea as their center of connectivity' (Malkin, 2011: 13). The Greek thalassocracy thus 'not only merely crosses the sea but uses the sea ... exercising some degree of control over the sea' (Abulafia, 2014: 139). Through considering how this worked, we can reimagine the sea not as a geography that connects land, but rather as one that holds connectivity within it. Ways of enacting this sea in antiquity depended on the employment of multiple methods of seafaring. This included thalassic networks that reflected a manipulation of distance and dynamic nodal systems that included the use of geographical features such as islands. This reveals how power and order were realised *on* and *through* the sea, rather than more grounded geographies.

Thalassic networks

Towns have traditionally been interpreted as central to the navigational routes of the Mediterranean in antiquity, generating an understanding of 'sharply defined lines of communication and redistribution' through the Mediterranean (Horden and Purcell, 2000: 123). In consequence, the connectivity of the ancient Mediterranean Sea is often analysed through the prospect of major routes and lanes of traffic linking these distant geographies. However, the enactment of this sea during antiquity was far more heterogenous and reliant upon intricate connectivities that functioned at sea. Both Malkin and Horden and Purcell's work generates messy, difficult to map engagements with this geography in antiquity. Through focusing on 'microregions' and the ways they were tied together through micro-voyaging, what emerges is a 'broader interpretative context' that reveals the significance of the multidirectional connectivity of this sea, as well as demonstrates a decentralised enactment of power at sea (Horden and Purcell, 2000: 123).[4]

A system of 'microdistance voyaging' (Malkin, 2011: 156) was definitional to the way the sea was made during antiquity. It was of great significance not only to the Greeks and their method of colonising this sea, but was also employed by the Phoenicians and Etruscans, who were maritime trading communities of the Middle East and Italy, respectively. These journeys by smaller vessels, while still ultimately culminating in long distances travelled, were premised upon a nodal system, in other words, points where these boats stopped. As such, these journeys often adhered closely to coasts. This system of 'short distance coastal navigation' that the Greeks developed was termed *cabotage*, which explains 'long-distance trading done via short hops and changing agents' (Malkin, 2011: 49). Hence, beyond the major open sea journeys that demonstrate diversified channels, *cabotage* journeys reflected a multiplicity of nodes as well as methods of navigating this thalassic space. In fact, so networked and complex were these journeys, with 'myriad possible combinations of port, shelter, detour' that they were 'too various in direction ever to be fully recorded' (Horden and Purcell, 2000: 140). In consequence, Malkin (2011: 19) argues that maps fail to encompass such interconnectivity: 'Two dimensional representations of connectivity mostly turn out to be messy "spaghetti monsters" ... Missing are the network's multidirectionality, multidimensionality, and multitemporality.'

These vessels engaging in systematic coastal navigation, while small in size and in their capacity to carry cargo, were not insignificant in defining trade within this region and were likely responsible for more of the transport of goods and people than the larger networked voyages. They carried cargo as well as travellers and pilgrims, and functioned in trade and in piracy (Horden and Purcell, 2000: 140). These were hence not journeys that paled in comparison to longer journeys across deeper waters, for while they may not reflect a feat in engineering, they demonstrate a specific form of connectivity within this geography. Coastal navigation reveals a way of making the sea in which it not only becomes a central point of orientation; it also becomes defined by innovation and shifting links. The sea in this context is not a space between territories or ports, but rather a full space facilitating intricate nodal links within its watery body.

Emerging from a networked engagement of multidirectional connectivity at sea is a method of navigation reliant upon a dynamic nodal system. Reflecting this, Malkin (2011: 157) illustrates the system of ports in antiquity to be premised upon 'a "many-to-many" network that radiated from each node to the Mediterranean at large'. The web of islands in the Mediterranean allowed journeys to be shortened, as well as providing visible destinations. So important were islands to the nodal structure of the Mediterranean in antiquity that the Greeks determined a word for sections of land that lay opposite islands, this being *peraiai*. This term itself, in being oriented towards the mainland's relationship to an island and not the inverse, 'reflects the conceptual primacy of the maritime world' (Horden and Purcell, 2000: 133). Hence, within this heavily networked system, there was a shared understanding of the ocean as the centre of existence facilitating multidirectional interactions between a multiplicity of nodes (Horden and Purcell, 2000: 157).

In a multitudinous way, modern migration in the Mediterranean and the European efforts to limit migratory movements emulates ways of making this geography that were present in antiquity. This can be seen in the increased reliance on coastal navigation as well as the development of complex nodal systems at sea that have emerged as direct and indirect consequences of EU bordering policies. Due to the proximity of Eunavfor Med to Libyan waters and the training of the LCG, there is a higher interception rate of migrant vessels closer to the littoral geography of Libya. This proximity in interception also functioned alongside phase two of the Eunavfor Med programme, which stipulated the destruction of migrant vessels once migrants have been disembarked. In May 2018, Eunavfor Med reported that they had 'neutralised' 545 vessels (Council of the European Union, 2018). In anticipation of greater levels of interception of vessels and the subsequent destruction of boats, smugglers have since further limited the technology they afford migrant boats. Amnesty International (2017: 12) have identified a growing 'absence of a satellite phone', in addition to insufficient levels of fuel in migrant boats departing Libyan shores. As a result of Eunavfor Med's policy of destruction, smugglers also now use disposable, inflatable vessels that are less resilient to overcrowding and rough seas (Amnesty International, 2017: 12). The greater precariousness of migrant journeys inspired by Eunavfor Med has stimulated a growing reliance upon a form of coastal navigation. This is demonstrated not only in interceptions that occur in proximity of Libyan shores, but also in the disasters that transpire within this geography. Data from IOM indicate that 'shipwrecks between 16 June and 31 July [2018] took place well within 50 nautical miles from Libya's shores' (Villa et al., 2018). As such, these voyages are not only deeply unmappable and inconsistent, premised upon the shifting strategies of smugglers, they are also often only capable of micro-distant voyaging. Mobility regulation and its effects on migrant journeys are cultivating the re-emergence of short journeys in the Mediterranean.

Interceptions at sea function as nodes in a migratory journey, not only interrupting voyages, but changing their course. Migrants may be rescued or intercepted, transferred onto larger vessels or have their vessel towed. Whether migrants are taken to Europe or returned to a third state or the state where they embarked their journey, from the point of interception at sea, migrant journeys take different routes. While interceptions may be orchestrated by European coastguards or European programmes of migration regulation, NGOs, or private and commercial vessels, the point of interception reflects the commencement of a sorting of migrants, determining where migrants are disembarked and whether they have access to an asylum procedure (Fassin, 2012; Vaughan-Williams, 2015). As such, these inceptions at sea reflect entanglements which are deeply political, determining not only how and where a migrant moves, but whether they will have access to an asylum process.

While the regulation of migration has encouraged a shortening of the distances travelled in this sea, micro-distance journeying is also a technique employed in migrant journeys. This is demonstrated by the significance of islands within migratory routes in the Mediterranean. Routes to European islands from the southern Mediterranean coast are often far shorter, encouraging migrant journeys to depend upon island geographies within this sea. Italy's Sicily, Sardinia, and Lampedusa; Spain's Cuerta; Greece's Lesbos, Samos, Kos, Rhodes, and Chinos are all central landing points in Mediterranean migration routes (IOM, 2018: 19, 31, 24). Thus, in the context of migration and its control, there is a re-emergence of the navigational importance of islands.

Maritime migration is far more complex than routes between spaces of embarkation and disembarkation, demonstrating the sea to be a thick fabric defined by shifting nodes of connectivity. This imagining of the sea not only relocates a focus within migration studies to the making of the sea and its points of connectivity; it also contrasts contemporary scholarly work on maritime trade that focuses on the trends in the containerisation of shipping and the pendulum routes that link global ports with speed and efficiency (e.g. Medda and Carbonaro, 2007; Bevan et al., 2014). The sea here is not a space to be traversed, nor does it function as a backdrop to political events. Rather, it is emphasised as a dense geography made through varying navigational tactics. Through considering the 'omni-directional connectivity' (Horden and Purcell, 2000: 130) of the Mediterranean during antiquity, we begin to develop a language that pays attention to the messy and entangled way this sea is made and re-made, and how changing seafaring strategies reflect the political ordering of this sea.

Colonising thalassocracies

During antiquity, the Greeks had an authority in the Mediterranean that was both an effect of and a foundation to the navigational tactics they employed. Malkin (2011) refers to the influence they held over this sea as a thalassic colonialism. The term colonialism here is premised upon an identity of 'Hellenicity' that was sustained throughout these Grecian nodes as well as the establishment of the complex multidirectional interconnectivity that defined the Greek thalassocracy (Malkin, 2011: 210). The Greeks, having encompassed the Mediterranean by migrating throughout its coastal regions, developed a Panhellenism in this region that was not premised upon sustaining a homogenised culture, but was rather cultivated by tying together dispersed communities through the connectivity of the sea. Greeks 'dispersed without becoming atomized', allowing 'spatial distance and network dynamics' to become the 'virtual center' of their colonial vision (Malkin, 2011: 61). This exemplifies the centrality of the sea to the spatial imaginary of antiquity. While there was a diffusion of actors and operations in this sea, the sea was nonetheless a uniting space. The Greek term for the Mediterranean, *he heretera thalassa* or 'our sea', was informed by this interconnectivity, from which they developed a perspective of 'ship to shore' (Malkin, 2011: 216). This had the effect of inverting distance, 'places linked by sea are always "close" while neighbours on land may, in terms of interaction, be quite "distant"' (Horden and Purcell, 2000: 133). The engagement with the sea for the Greeks generated a perspective of the Mediterranean as a central space of political activity; the sea is not imagined as a space between grounded geographies, but rather a space crucial to the existence of hinterland communities. This is not to suggest the Greeks orchestrated a sea without contestation; their expansive colonisation within this region led to conflict and violence (see Abulafia, 2011: 132–49). Rather the aim is to emphasise that there were ways of using this sea, in concomitance to several religious and maritime cults, that 'bound together the emerging Athenian empire' (Abulafia, 2011: 138).

A broad view of this sea during antiquity, across the connected microregions linked by a multiplicity of routes, reveals a form of control premised upon a process of 'convergence through divergence' (Malkin, 2011: 5). In other words, authority for the ancient Greeks operating in this thalassic geography was realised through their deeply 'decentralized network' (Malkin, 2011: 8). It was a region whereby encounters between Greeks and non-Greeks were central to the formation of a Panhellenic identity, and where variant manifestations of a Greek identity across vast geographies were unified by intricate connections across this liquid space as well as a coherent imagination of the sea (Malkin, 2011: 87). This is illustrated by Plato, who describes the ancient Greek society as 'living around the sea like ants and frogs round a pond' (*Phaedo* 109B). While spatial distance and cultural difference were definitional to this region, what this era demonstrates is a governing at sea that was diffused, realised through seafaring strategies and forms of at-sea connectivity.

Similar to the Greek thalassocracy, the agenda of mobility regulation in the Mediterranean is reflected in the shifting seafaring strategies at sea. As was explored in the prior section, modern mobility control is reconfiguring navigational strategies at sea. Within these changing tactics, there emerges a distinct authority in this geography premised upon a 'convergence and divergence'. This is demonstrated by the way European actors are working with local authorities in southern Mediterranean states, such as in the relocation of a European programme to Libyan territorial waters through the training of the LCG. A European agenda is similarly being deployed to the Tunisian coast. In late January 2018, Tunisia signed a Cooperation Plan with Italy. Italy's former Defence Minister, Roberta Pinotti, indicated that the Plan would include multiple activities that would assist in strengthening the 'military and security operations' at sea (Ministry of Defence, 2018). This 'convergence and divergence' is also exemplified in the broader influences of European policy on seafaring strategies. This is exemplified by a proliferation in shorter modes of travel at sea, while the disincentivisation of migrant rescue can equally be interpreted as the effect of a European agenda on the way this sea is made (Basaran, 2015; Stierl, 2016). The diffusion of various agents monitoring and assisting the regulation of mobility at sea inscribes not a Panhellenic identity, but instead a European agenda of migration management and exclusion.

Imagining the Mediterranean Sea through the lens of the Greek thalassocracy facilitates the development of a language that treats the sea as a central vantage point, through which we can consider the connectivity that shapes this sea and what it reveals of political agendas at sea. This requires that we prioritise an analysis of seafaring practices and at-sea actions. In examining contemporary maritime migration management, what we see is a complex web of engagement at sea that is ever-changing, the effect of which is intra-state and sea-wide. Through this, we can consider how various actors and authorities in this region are influenced by European policies of mobility control. These shifting forms of at-sea connectivity are not unique to the Mediterranean. In the context of migration management, various maritime geographies around the world reflect diffused logics of migration containment and the authority of state actors operating far beyond territorial waters to interdict migrants. In the US, the interdiction of migrants at sea proliferated in the early 1990s during a period of Haitian migration. During this time the US government not only normalised a practice of interdiction on the high seas, but also the detention of migrants on Coast Guard Cutters and the return of migrants via sea to sites of detention or their origin country. Equally in Australia, strategies of interdicting, detaining, and returning migrants at sea proliferated during the early 2000s. These contexts are equally not easily defined by maritime law or security practices. Rather, they evidence changing types of connectivity at sea, practices of interception and return that do not easily fit with principles of international law, and a political authority that influences seafaring strategies. There is a need to engage with these specific practices at sea and what they reveal to us of maritime geographies. Centralising the maritime geography as a space of politics facilitates a greater understanding of how power and authority operate through this geography. Through this, this sea emerges as much more than a geography to be traversed on the way to more terrestrial geographies; rather, it is configured as a dense political space, defined by the connectivities and entanglements within it.

Conclusion

The Mediterranean Sea has, for millennia, been determined by the connectivities and entanglements which transpire within it. Different eras in this sea's history could have been drawn upon to illustrate the significance of examining the at-sea practices which define this sea. Succeeding the Greek thalassocracy was that of the Roman's, yet they held a distinct view of the sea. Their version of '*mare nostrum*' or 'our sea' had Rome at its core, centrifugally influencing distant territories (Malkin, 2011: 5). This differed from the Greek version of a shared sea, which treated the Mediterranean as a central geography, generating a perspective deeply rooted in the thalassic. Following the Roman Empire was the Arabic Empire, which equally relied upon the sea, although a number of scholars have emphasised that the sea in this era was treated as 'a frontier rather than a unity and new non-maritime capitals became important' (Harris and Harris, 2005: 5). As a period of connectivity and change, the fifteenth century and Braudel's (1995) exposition on the Mediterranean could equally have been developed. Braudel demonstrates that the Mediterranean expanded far beyond the sea, through routes and networks from the Balkans to Sudan. As such, Braudel (1995: 276–7) argues that the Mediterranean needs to be considered as whole, as an ensemble. However, the purpose here is not to generate a *longue durée* of this sea, nor develop expansive comparisons between eras of this sea. The purpose is to begin a project of reimagining the sea as a central political geography in order to consider the sea, like its more terrestrial counterparts, as densely occupied and defined by the practices which transpire within it.

While the politicisation of maritime migration in the Mediterranean has transformed the sea into a space of paramount significance, the way this geography is being rendered as an effect of the enactment of European agendas of mobility control remains unconsidered. Through studying the practices in the central Mediterranean route by Eunavfor Med in this chapter, what is demonstrated is the fluid, dynamic, and entangled way migration regulation works in this sea. As the Eunavfor Med programme has drawn to a close, the discussion of how this sea is perpetually being redefined is more pertinent than ever. We are limited in our ability to consider the effect of these practices and policies on the making and re-making of this sea. Through borrowing on an imaginary of this sea from the ancient Greeks, we are able to emphasise the significance of connectivity at sea. Considering the Mediterranean through the Grecian thalassocracy emphasises how this sea is far more than a liminal geography to be traversed on the way to more terrestrial landscapes. Rather, the Greek thalassocracy illustrates a thalassic perspective which takes seriously the connectivities and entanglements that shape this geography. As such, this chapter proposes a language for the sea that not only emphasises the various and changing forms of at-sea connectivity, but considers what they reflect of a political power in maritime geographies. The contemporary Mediterranean Sea is a complex geography of activity, yet it is one in which an objective of interrupting migration from the southern and eastern shores of the sea increasingly influences the seafaring strategies and forms of connectivity. Reading the sea through connectivity and entanglement opens up a way to interrogate how migration management influences the broader conditions of this sea, as well as how seafaring is tied in with political ordering.

Notes

1 Of the 21 littoral states of the Mediterranean, claims of a contiguous zone have only been made by Algeria, Cyprus, Egypt, France, Italy Malta, Monaco, Morocco, Spain, Syria and Tunisia (Papanicolopulu, 2015: 608). Only seven states have claimed an EEZ: Morocco, Syria, Cyprus, Tunisia, Libya, Lebanon and France. There is some ambiguity surrounding Egypt and Israel (Papanicolopulu, 2015: 608).

2 Article 8.7 stipulates that: 'If evidence confirming the suspicion [of smuggling] is found, that State Party shall take appropriate measures in accordance with relevant domestic and international law.'
3 For a social political history of this sea, see Braudel (2002), Chambers (2008) and Harris and Harris (2005).
4 This is not to suggest that major routes of transport and trade in antiquity were inconsequential or overstated.

References

Abulafia, D. (2011). *The Great Sea: A Human History of the Mediterranean*, Oxford: Oxford University Press.

Abulafia, D. (2014). 'Thalassocracies', in P. Horde and S. Kinoshita, eds, *A Companion to Mediterranean History*, London: Wiley, 137–53.

Agnew, J. (1994). 'The territorial trap: the geographical assumptions of International Relations theory', *Review of International Political Economy* 1(1): 53–80.

Amnesty International (2017). *A Perfect Storm: The Failure of European Policies in the Central Mediterranean*, London: Amnesty International.

Basaran, T. (2010). *Security, Law and Borders: At the Limits of Liberties*, Abingdon: Routledge.

Basaran, T. (2015). 'The saved and the drowned: governing indifference in the name of security', *Security Dialogue* 46(3): 205–20.

Bevan, A., R. Greenberg, C. Knappett, M. Lawall, A. Meneley, N. Purcell, E. S. Sherratt, D. L. Smail and A. Bevan (2014). 'Mediterranean containerization', *Current Anthropology* 55(4): 387–418.

Bevilacqua, G. (2017). 'Exploring the ambiguity of Operation Sophia between military and search and rescue activities', in G. Andreone, ed., *The Future of the Law of the Sea*, Cham: Springer Nature, 165–89.

Biersteker, T., and C. Weber (1996). 'The social construction of state territory', in T. Biersteker and C. Weber, eds, *State Sovereignty as Social Construct*, Cambridge: Cambridge University Press, 1–21.

Braudel, F. (1995). *The Mediterranean and the Mediterranean World in the Age of Philip II*, vol. 2, Berkeley, CA: University of California Press.

Bueger, C. (2015). 'What is maritime security?' *Marine Policy* 53: 159–64.

Bueger, C., and T. Edmunds (2017). 'Beyond seablindness: a new agenda for maritime security studies', *International Affairs* 93(6): 1293–311.

Cacaud, P. (2005). *General Fisheries Commission for the Mediterranean Fisheries Laws and Regulations in the Mediterranean: A Comparative Study*, Rome: FAO.

Carrera, S., and L. Den Hertog (2015). *Whose Mare? Rule of Law Challenges in the Field of European Border Surveillance in the Mediterranean*, CEPS Liberty and Security in Europe No. 79.

Chambers, I. (2008). *Mediterranean Crossings: The Politics of an Interrupted Modernity*, Durham, NC: Duke University Press.

Council of the European Union (2016). *Eunavfor Med Operation Sophia Authorised to Start Two Additional Supporting Tasks*, Brussels, 30 August.

Council of the European Union (2017). *Malta Declaration by the Members of the European Council on the External Aspects of Migration: Addressing the Central Mediterranean Route*, Brussels.

Council of the European Union (2018). *EUNAVFOR MED Operation Sophia: Operation to Contribute to Better Information Sharing on Crime in the Mediterranean*, Press Release. 14 May.

Council of the European Union (2019). 'Infographic – EU Mediterranean operations 2015–2019', www.consilium.europa.eu/en/infographics/saving-lives-sea-february-2018/ (accessed 24 January 2020).

Cusumano, E., and J. Pattison (2018). 'The non-governmental provision of search and rescue in the Mediterranean and the abdication of state responsibility', *Cambridge Review of International Affairs* 31(1): 53–75.

Cuttitta, P.. (2018). 'Pushing migrants back to Libya, persecuting rescue NGOs: the end of the humanitarian turn (Part I)', *Border Criminologies*, www.law.ox.ac.uk/research-subject-groups/centre-criminology/centreborder-criminologies/blog/2018/04/pushing-migrants (accessed 2 April 2019).

European Council (2015). 'Operation Sophia: Fighting human trafficking in the Mediterranean', https://eur-lex.europa.eu/legal-content/EN/TXT/HTML/?uri=LEGISSUM:4309220&rid=4 (accessed 10 February 2022).

European External Action Service (2018). *Strategic Review on EUNAVFOR MED Op Sophia, EUBAM Libya and EU Liaison and Planning Cell*, 27 July, Brussels.

Fassin, D. (2012). *Humanitarian Reason: A Moral History of the Present*, trans, Berkeley, CA: University of California Press.

Harris, W. V., and W. V. Harris (2005). *Rethinking the Mediterranean*, Oxford: Oxford University Press on Demand.

Heller, C., and L. Pezzani (2012). *Forensic Oceanography: Report on the 'Left-To-Die Boat'*, https://forensic-architecture.org/investigation/the-left-to-die-boat (accessed 10 February 2022).

Horden, P., and N. Purcell (2000). *The Corrupting Sea: A Study of Mediterranean History*, Oxford: Blackwell.

Hyndman, J., and A. Mountz (2008). 'Another brick in the wall? Neo-*refoulement* and the externalization of asylum by Australia and Europe', *Government and Opposition* 43(2): 249–69.

International Organization for Migration (IOM) (2018). *Mixed Migration Flows in the Mediterranean* (Report), March, https://reliefweb.int/sites/reliefweb.int/files/resources/Flows_Compilation_Report_March_2018_0.pdf (accessed 10 February 2022).

Klein, N. (2011). *Maritime Security and the Law of the Sea*, Oxford: Oxford University Press.

Klein, N. (2015). 'Maritime security', in D. R. Rothwell, A. G. Oude Elferink, K. N. Scott and T. Stephens, eds, *The Oxford Handbook of the Law of the Sea*, Oxford: Oxford University Press, 582–603.

Malkin, I. (2011). *A Small Greek World: Networks in the Ancient Mediterranean*, Oxford: Oxford University Press.

Medda, F., and G. Carbonaro (2007). 'Growth of container seaborne traffic in the Mediterranean Basin: outlook and policy implications for port development', *Transport Reviews* 27(5): 573–87.

Ministry of Defence (2018). 'Ministry of Defence: Italy-Tunisia Cooperation Plan signed', www.difesa.it/EN/Primo_Piano/Pagine/tunis.aspx (accessed 10 February 2022).

NATO (2016). *NATO Summit Guide*. Warsaw, 8–9 July 2016, www.nato.int/nato_static_fl2014/assets/pdf/pdf_2016_07/20160715_1607-warsaw-summit-guide_2016_eng.pdf (accessed 10 February 2022).

Papanicolopulu, I. (2015). 'The Mediterranean Sea', in D. R. Rothwell, A. G. Oude Elferink, K. N. Scott and T. Stephens, eds, *The Oxford Handbook of the Law of the Sea*, Oxford: Oxford University Press, 604–26.

Papastavridis, E. (2008). 'Interception of human beings on the high seas: a contemporary analysis under international law', *Syracuse Journal of International Law and Commerce* 36: 145–227.

Papastavridis, E. (2016). 'EUNAVFOR Operation Sophia and the international law of the sea', *Maritime Safety and Security Law Journal* 2: 57–63.

Ryan, B. J. (2015). 'Security spheres: a phenomenology of maritime spatial practices', *Security Dialogue* 46(6): 568–84.

Shah, N. (2012). 'The territorial trap of the territorial trap: global transformation and the problem of the state's two territories', *International Political Sociology* 6(1): 57–76.

Steinberg, P., and K. Peters (2015). 'Wet ontologies, fluid spaces: giving depth to volume through oceanic thinking', *Environment and Planning D: Society and Space* 33(2): 247–64.

Stierl, M. (2016). 'A sea of struggle – activist border interventions in the Mediterranean Sea', *Citizenship Studies* 20(5): 561–78.

UNHCR (2018). *UNHCR Flash Update, 20–26 January 2018: Libya*, https://reliefweb.int/sites/reliefweb.int/files/resources/UNHCR%20Li bya%20Flash%20Update%2026%20January%202018.pdf (accessed 10 February 2022).

Vaughan-Williams, N. (2015). *Europe's Border Crisis: Biopolitical Security and Beyond*, Oxford: Oxford University Press, USA.

Villa, M., R. Gruitjers and E. Steinhilper (2018). 'Outsourcing European border control: recent trends in departures, deaths and search and rescue activities in the Central Mediterranean', www.law.ox.ac.uk/ research-subject-groups/centre-criminology/centreborder-criminologies/ blog/2018/09/outsourcing (accessed 18 September 2018).

Williams, J. (2015). 'From humanitarian exceptionalism to contingent care: care and enforcement at the humanitarian border', *Political Geography* 47: 11–20.

7

Constructing insecure maritime spaces: navigational technologies and the experience of the modern mariner

Jessica K. Simonds

There's anxiety in putting up the razor wire and getting ready, it starts to get real, then we just cross a line and it's imaginary.
— Officer Madeleine Wolczko, 9 March 2019

This chapter explores how maritime security resources that embody politically charged discourses such as maps, and practical security guides are carried to sea. Through global seafarer engagement, these resources frame the consciousness behind the navigation of risk through the construction of insecure spaces, which provoke exceptional practice in the pursuit of safe transit. Merchant navy seafarers engage with these technologies to navigate the seas that consume 71 per cent of the world's surface (Anderson and Peters, 2014: 3), to bring us over 90 per cent of everything we consume (Lansing and Petersen, 2011: 513). In the broader sense, this chapter will address questions such as *how do we navigate the sea? How is it tamed in our imaginations? And how do we consider threats to our marine transit a threat to our landed political structures?* In response, I propose that we navigate risks at sea by drawing on uncomfortable imaginations of transiting different spaces, many of which are displayed on maps and within the tools we are presented with to navigate the map as a representation of the world. Through conforming to the imageries represented on maps through navigational practice, insecure social and physical spaces are constructed on board the site of the merchant vessel. This will be addressed in various historical scenarios, but specifically by drawing on empirical reflections of seafarers who have transited the Horn of Africa at the peril of contemporary piracy.

In bringing the sea back into International Relations (IR), more specifically critical security studies, this process perpetuates the production and reproduction of what the so-called Paris School scholars' term 'fields of insecurity' (Bigo and Tsoukala, 2008: 8). Yet, combined with the seas' materiality and vessel mobility, this is imagined as a dynamic and transformative process where the construction and deconstruction of these fields is built into the navigational planning of the mariner. Security is defined as a product of performance, orchestrated by professionals of politics who cultivate social spaces defined by unease and anxiety (Huysmans, 2006; Huysmans et al., 2006; Bigo and Tsoukala, 2008; Bigo, 2016) in response to a variety of bureaucratically managed (Van Munster, 2007) existential threats. This security framework has been applied to immigration centres (Huysmans, 2006: 55), airports (Salter, 2008) as well as to the developmental challenges in constructing threats pertaining to Somali piracy (Gilmer, 2019). Yet, the temporary transformative capacity of security sites in both the physical and social sense has not been explored in relation to the mobility of the merchant vessel, nor has the agency of the seafarer. Taking stock of its propulsion at sea and in reflection on the opening quote of this chapter, the seafarer both constructs as a producer, and experiences as a consumer, the field of insecurity based on engagement with navigational technologies. Therefore, focusing on the merchant vessel as a social and physical security site, the transformation in transit is apparent in planning passage and designating exceptional security roles at sea. This chapter will draw on a historical range of examples to demonstrate that maps and navigational guides are discursively rich resources. In bringing the sea *back* in IR, they serve to symbolise how competitive political discourses frame how we look out to sea and envision its geography as a utility of landed politics.

The scope of this chapter does not allow me to address land-based studies that deter the root causes of Somali piracy (Boren, 2014; Keenan, 2014; Ivancovich, 2015), the specifics of the trade consequences that are realised on land (Cummins and Trainar, 2009; Blomberg et al., 2013; Germond, 2015) or the exclusive process of shore-based competition over the authority of counterpiracy guidance (Hansen, 2012; Bueger, 2013, 2018). It will focus on the maps and practice guides as techniques of governing the navigation of risk. Navigational charts, instruments and practical advice will here be understood in a broader sense, as techniques of governance in a Foucauldian sense (Patton, 2013: 179). In this perspective, it is evident that modern maritime security technologies have been designed to protect the timely transit of trade, through procuring specific modes of navigational practice into the life of the seafarer. Counter-piracy guidance prioritises denying a potential pirate access to the vessel as a method of mitigating the opportunity for a full hijacking, which would delay the cargo, the crew as well as the vessel – proving expensive for all who have a stake in the journey. Modern cargo ships such as the container ship have revolutionised the capacity of merchant shipping (Levinson, 2006) as they have been designed to architecturally prioritise the capacity of the vessel as opposed to defensive security. Hence, the construction of insecurity at sea is based on the agency of the crew to both socially and physically transform the vessel into a space that is prepared for a potential attack. As the empirical section of this chapter will demonstrate, the dynamism in how responses to maritime threats have been realised calls for an open methodological approach in navigating the future of ventures at sea.

Conceptual approach

Traditional Realist and Liberal approaches to addressing the sea both encompass a state-centric perception of security, centred around the pursuit of relative or absolute gains of power (Walt, 1998). Therefore, the majority of traditional scholarship pertaining to the seas has substantiated the oceans as a 'battleground' (Steinberg, 2001) where the pawns of the state (navies) engage in wargames as an extension of landed politics. Access to and control over specific spaces such as the Black Sea and the Straits of Hormuz are integral to the hegemonic struggles between the United States and its competitors, be they Russia, China or Iran. Nevertheless, command of the seas is often considered a pre-requisite for global hegemony, and in the twenty-first century, the United States is considered to be the dominant naval power. This oceanic hegemonic position has largely allowed it to do what it pleases at sea

(De Nevers, 2015). The sea, as such, is seen as a backdrop to the game of battleships, and imagined as a friction-free surface (Steinberg, 2001: 165) for the games of war and power to take place over spaces of control and exclusivity, rather than addressing the sea's innate material differences. This in turn would require a more open approach to assessing how we interact, navigate, propel and tame the sea in these endeavours. By pairing a critical security framework with a 'wet ontological' (Steinberg and Peters, 2015) perspective, the layered materiality of the sea that positions the merchant vessel as a floating and mobile infrastructure can be accounted for along-side an institutionalised engagement with security. This simultan-eously broadens our imaginations beyond the state as an exclusive security actor and deepens our perception of why technologies such as maps and practical guides are so important in constructing and navigating the materiality of the sea.

The application of a practice-oriented conceptualisation of inse-curity that considers the challenges and benefits of the seas' wet and layered materiality allows us to draw on security as a performa-tive, mobile and individualised experience, which takes account of the seafarers' skilful yet vulnerable relationship with life at sea. Scholars such as Didier Bigo have produced an 'international polit-ical sociology' (Bigo, 2016) of security, where it is not only viewed through the lens of the professionals of security, or in a state-centric context, but through engaging with questions such as 'how security is understood, articulated and experienced?' (Bigo, 2016: 1081) by individuals in the everyday. Security can therefore be experienced by anyone and orchestrated by a 'web of competitive relations among security agencies, government officials and private com-panies' (Bourne, 2013: 59). At sea, these have constructed a heavily regulated normative regime of behaviours for seafarers to both sur-vive and thrive. Because of this individualised approach, otherwise non-political or security-oriented actors are identified as significant security actors with power to direct behaviour and construct risky spaces based on their bespoke expert knowledge of shipping and seafaring. An example that will later be addressed is the role of Protection and Indemnity (P&I) Clubs as insurance groups. While these do not necessarily 'pick up the big bills' (Simon, personal correspondence, 2018) of maritime risks such as piracy, they have nevertheless drawn on their outreach capabilities and safety expertise to contribute to the drafting and dissemination of counter-piracy guidance. This approach posits security as an energetic concept, with the potential for revision in its governance by a myriad of security professionals, who are neither statically positioned in the field nor explicit in their contribution to a discourse of security.

I use the concept of field in Bourdieu's sense as an autonomous social space where actions, interpretations and translations take place (Bourne, 2013: 60). Flick argues that hardly any institutional activity occurs without producing a record (Flick, 2018: 375) and therefore maps and practical guides serve as a politically charged record of how maritime space has been envisioned by powerful landed institutions. Through engagement with the seafarer, insecure space is produced both socially and physically through the implementation of an exceptional regime. This space is produced when a vessel is preparing to enter, and in navigating, the High Risk Area (HRA) as a geographical space designated on the Q6099 Maritime Security Chart and is realised through environmental feelings of anxiety, unease and apprehension. It is physically realised through the architectural transformation of the vessel with razor wire, hoses and navigational manoeuvrability. As such, these documents can be analysed in their construction of transformative security spaces and affirmed by the experiences of the seafarer. They form the link between land-based politics and maritime seafaring practice.

The catalyst for producing insecure space is the imagination of a potential encounter with a predefined existential threat (Bigo, 2016: 1078) that may or may not manifest. The construction of this threat is not through the mechanism of a 'productive moment' (Wæver, 2011: 468), or the consequence of an elite 'speech-act' (Bourne, 2013: 59) that the Copenhagen School presents as a response to a grand turning event. But through embedding exceptional practices into daily life, this leads, through human engagement, to the self-production of norms in specialised scenarios. These are visible as behavioural regulations that mitigate the *possibility* of a disastrous encounter or presence of an undesirable *other* that threatens the success of the voyage. The actors with a capacity to direct the consciousness as to what constitutes a viable threat hold power in their construction of insecure spaces on maps and within practice guides – and must be able to convince the seafarer that the threat is viable for them to comply with the spatial vision they are given.

The partnership of maps and practical guides is important in understanding the production of insecure space as the map represents where is considered risky (based on an imagination of risk) and the practical guide provides the behavioural toolkit for navigating the sea safely based on this imagination of risk. Drawing on Lefebvre's spatial conceptualisation, social space can be imagined as a dynamic process, in a constant state of familiar rhythmic performance (Lefebvre, 2004: 42). Through engaging with both the map and the practical guide as a toolkit, the seafarer legitimises the construction of insecure space through performative practice and conformity. The dynamism of this process is defined by differing perceptions of what constitutes risk and what should be included within the practical guide as a toolkit in which to navigate it. Somali-based piracy, for instance, became a big business rapidly between 2008 and 2011, leading to the release of multiple practice guides as the threat evolved. This process is continuous and responsive to both experiences of risk at sea and how they correlate to the interests of shore-based political regimes. Therefore, the administration of an insecure spatiality and the endurance of control of the sea is managed by a rhythmical relationship and communication process from land to sea. As a constant process of resource revision, communication, seafarer performance and feedback, this characterises the spatial production process as both dynamic and open in the consciousness of what is considered the risk that insecure space is framed upon.

The map and its tools: entangling representation and navigational technologies

Maps offer visual representations of places by embodying an individual or collective mental vision of space (Lefebvre, 1974) and are the first navigational tool this chapter will address in its embodiment of institutionally constructed spatial visions. Maps display political discourses, philosophies and familiar frames of reference for the navigator to translate into methods of practice. The power behind the map is therefore consolidated in its capacity to make 'statements about the world' (Wood, 1992) through projecting a perspective of how terrestrial and hydrographic space is ordered.

Where a place is marked as dangerous or risky on a map, the provocation of cautionary behaviours legitimises this designation and thus the construction of insecure space. Methods of which have historically served to 'derive the idea of property' (Netz, 2004: xi) and project connotations of inclusion, exclusion, containment and authority to legitimise and project territorial claims in order to shape the way we behave in and between places.

Maps are loaded with ideologies that are reflected through presentation, scale, technological language and use of colour in their visual representation of risk-averse areas. For seafarers, the map takes a position at the centre of the bridge, where paper maps still are the basis to inform of the passage plan. Engagement with the map as a matter of daily routine makes them an important resource in governing insecurity, as the sea is carved into territories that are spatially legitimised through alterations in practice while in transit. Without behavioural compliance and enforcement, there are no infrastructures such as walls or fences to define them. In the modern context, mapped risky maritime areas such as the Horn of Africa, the Malacca Straits, West Africa, and the Mediterranean Sea have become associated with threats such as armed robbery, piracy, drug trafficking and illegal migration. As such, specific track lines, risky areas as well as aversion zones, are designed on maps that are used to transit them. This is a practice that is not only evident in the modern context, but throughout histories of engagement with the sea where a threat or dangerous area has been defined by a powerful shore-based regime which then seeks to assist specific seafarers in their pursuit of safe transit based on their relationship with them or their endeavour.

Ancient imaginations of mapping and navigating risk

Ancient conceptualisations of maritime risks have been represented and embedded within navigation practices and instruments and offer an example as to how threats are situated and vilified in their relation to landed societal politics and imaginations of danger. The performance of exceptional navigational practices is part of the loop of spatial production and captures the compliance of the seafarer based on the presentation of a credible danger. Linebaugh and

Rediker define the construction of a maritime danger as the meta-phorical 'many-headed hydra' (Linebaugh and Rediker, 2012: 3). As a blanket term for threats and challenges that emerge from the sea to instil fear in those who traverse the waters, the hydra draws on landed imaginations of the very worst fear for the mariner, often drawing on dangers to the mission of the voyage or purpose of maritime transit as well as life and prosperity. In the modern era, the hydra can be reimagined as an 'antithetical symbol of disorder and resistance, a powerful threat to the building of state, empire, and capitalism' (Linebaugh and Rediker, 2012: 3) and encompasses threats such as piracy, drug smuggling, the trafficking of stowaways as well as illegal fishing and maritime terrorism. Historically, the hydra can be imagined as any danger that defines spaces of unfamiliarity, caution and unease, as experienced in Babylon and stories deriving from Ancient Greece and Rome.

The variety of maps that reflects the ordering of the world reflects that there is *no true map* – only subjective maps that have been drawn to represent space as it is relevant to our own frames of political reference, drawing on familiar societal factors and imagin-ations of the unfamiliar. A clay tablet depicting a map of Babylon (see the British Museum, Map of the World, 92687) originated around the first millennium BCE (Verderame, 2020: 83) and is considered the first map of the world. The map depicts Babylon and the Euphrates river as central, reflective of how Babylonian civilisation was considered the centre of their worldly imagination – distant societies had not yet been *discovered* and were therefore positioned as uncomfortable and dangerous spaces. As the distance increases from Babylon, so does the ambiguity and mystery that surrounds the cartographic representations on the map. Peripheral to the familiarity of Babylon, cities such as Assyria and Susa are depicted further away, represented by smaller, less detailed circles. The sea is depicted in its ambivalence to Babylonian society, where it is referred to as a 'bitter river' that separates the land of the living from that of the dead (Verderame, 2020: 83) and an example of how geography has been visually displayed to separate sociological aspects of life and death. The map is animated in Babylonian lit-erature, where the personification of the water as a hostile and anguished materiality is depicted in the Epic of Gilgamesh, where a deluge destroys life and seals the fate of many to the bottom of the sea. As a dangerous and troublesome materiality, the sea is weaponised in this myth. The Gods bring rains and floods that 'tear against the sky and the land … and against the people like an army', bringing 'rage … darkness and noise' (Dalley, 2000: 31). Drawing ocean imaginations and death as intertwined concepts, the powerful Gods are the only entities with the capacity to tame and orchestrate safe passage. As a consequence, far-flung maritime endeavours were considered dangerous, voyages risky, and the sea *untamed* based on both these mythical depictions against cartographic representations of the sea.

The sun and birds are also embedded in technologies and tools of navigation and reflect how landed familiar frames of reference are utilised to distinguish unease based on their absence. In the *Epic of Gilgamesh* and *The Flood*, the sun and sea hold important connotations of life and prosperity that are also represented on the Map of the World. The sun draws on symbolism of healing, generosity and protection (Woods, 2012: 84–5), where the setting of the sun reflects the transition between life and death – security and danger, with the space between being the uneasy insecure spatiality where the unknown of death provokes anxiety. In the Sumerian narrative of *The Flood*, Ziusudra as the protagonist and hero is settled 'in the land Dilmun, where the sun rises' (Verderame, 2020: 80), which is a signifier for his future prosperity. The *Temple of Hymns* composition describes the Goddess Nanše as 'born on the shore of the sea, who laughs on the foam of the sea, who plays on [the water] of the waves' (Verderame, 2020: 85). Nanše holds patronage over the nourishment and care over the fish and birds, delineating an origin for these creatures. As such, the 'Map of the World' draws on the sun and birds as a measurement in which spatial familiarity and in turn security can be sought. Insecurity is framed by their absence and peripheral islands are referred to as 'a place where the sun cannot be seen' and beyond the flight of birds' (Andrews, 2016, n.p.; compare with Danzer, 1991) and are therefore positioned beyond familiar and comfortable points of reference for those who navigate beyond the map's familiarity. The representation of these islands reflects the scale in which society was measured by drawing on life and cosmos as the points of reference where life at sea could be imagined – and most importantly was considered safe. Alluding to the absence of prosperity and life on the map is therefore an early example of cartographic practice that constructed spaces of insecurity by drawing on shore-based political ideologies and aspects of familiarity in the construction of danger, a practice that is mobilised in ancient civilisations and their chronicles of navigating risk.

Navigating specific recurring dangers and risks are also evident in Ancient Greek and Roman imaginations of the hydra; these examples reflect not only the seas' materiality as represented in maps but the agency of the mariner in calling on guidance to transit the sea. As polytheistic civilisations, assistance and guidance from powerful actors such as the Gods and Goddess of Greece and Rome were integral for the successful transit of risky maritime spaces. Protagonist heroes depicted in Homer's *Odyssey*, Virgil's *Aeneid* and Apollonius Rhodius' *The Argonautica* evidence how the assistance of the *powerful* can lead to a level of comfort and success in navigating a peril of the sea. Charybdis is one of the dangers that characterises the Strait of Messina as a dangerous maritime chokepoint throughout these poems. She manifests herself as a violent whirlpool that 'gulps the dark water down. Three times a day she vomits it up, three times a day she gulps it down' (Homer, 2002: 288), invocative of imageries of a volatile and turbulent beast. On the other side of the strait, Scylla poses her own horrific manifestation as 'a grisly monster, I assure you, no one could look at her with any joy, not even a God who meets her face-to-face … she has twelve legs, all writhing, dangling down, and six long swaying necks, a hideous head on each' (Homer, 2002: 289). With each head, Scylla 'snatches up men from the dark-prowled craft and whisks them off' (Homer, 2002: 289). In each of the ancient poems it is explained that Scylla is undefeatable and therefore only escapable through performing specific navigational manoeuvres and careful planning with the assistance of the powerful Gods that will determine each hero's ability to escape the threat she poses.

The story of Odysseus' ten-year journey home from the Battle of Troy that is detailed in Homer's *Odyssey* can serve as a travel guide to the Mediterranean Sea through following his navigational journey. Odysseus does not have the most powerful Gods on his side in navigating the risky Straits of Messina and must rely on the Goddess Circe, who is depicted as an outcast and villain. She suggests he remains on the side of Scylla, by drawing on the logic that it is 'better by far to lose six men and keep your ship, than lose your entire crew' (Homer, 2002: 289). Odysseus follows this advice and loses men – yet can escape by following her navigational advice. Aeneas within *The Aeneid* encounters Scylla and Charybdis on multiple occasions and is advised by Andromachë and Helenus in his navigational planning. They suggest to 'waste time, skirting Sicily in a long arc rounding Cape Pachynus, than once set eyes on gruesome Scylla deep in her cave' (Virgil, 2006: 112). This approach still constructs an insecure space by drawing on the Straits of Messina as dangerous, and the exceptional navigational approach is defined by a decision whereby the time added to the journey is a more comprehensible inconvenience than the loss of men or the vessel entirely. Also, Jason and his Argonauts employed the assistance of the goddesses Hera (wife of the powerful Zeus) and Athena (Goddess of War) to guide him in his pursuit of the Golden Fleece, as depicted in *The Argonautica*. With their assistance, 'Thetis [Goddess of Water] set her hand to the rudder-blade, to guide them amid the Wandering rocks, and as when in fair weather herds of dolphins come up from the depths … guided its course' (Apollonius of Rhodes, 1998: 120). Jason's relationship with the more powerful goddesses provided him with a navigational advantage over Aeneas and Odysseus. With the assistance of Thetis, who drew on her specialised expertise of taming the sea, his risks were mitigated in his escape, which ultimately provided him with safe passage. In the modern context, the assistance of powerful shore-based polities and industrial actors also provides an advantage to merchant vessels who have access to specific technologies, naval assistance as well as access to specialised maps and practice guides to navigate modern perils. Therefore, the ancient examples demonstrate that the success of practices and performances of maritime security are entangled with landed political regimes, their assistance and their capabilities.

This section has demonstrated ancient examples of mapping and navigating risky maritime spaces. The Babylonian example of cartographically representing space has embodied the sun, sea and birds as familiar points of reference in the construction of both safe and insecure navigational endeavours. The ancient polytheistic examples of powerful Gods facilitating the navigation of risk reflect that security at sea does not happen within a vacuum, but is based on a fluid and dynamic relationship based on expertise, guidance and planning when making decisions to transit danger. Although the historical risks that have been addressed may seem detached from modern navigation in dangerous waters, these concepts are evident in constructing and planning navigating risk and insecurity today – in transforming a vessel into a space of insecurity in the deterrence of Somali-based piracy.

Mapping and navigating contemporary piratical waters

The Internationally Recognised Transit Corridor (IRTC) was established in 2008 as a collaboration of a multinational conglomeration of naval patrols including European Union (EU) and NATO missions. It was the first delineation of an exceptional maritime space whereby specific navigational practices were required to deter piracy. These patrols facilitated convoys of merchant vessels from the chokepoint of the Bab-Al-Mandeb, through the Gulf of Aden as a narrow strait between the northern coast of Somalia and the southern coast of Yemen. The corridor consisted of two one-way track lines, one facilitating vessels travelling east and the other west, both supported by naval patrols who can maximise their presence at a single designated route. Cowen argues that the IRTC was 'literally the production of a new political space, since it established new forms of authority and legal regulations' (Cowen, 2014: 153 as cited in Bueger, 2018: 624) outside the realm of normative maritime law and practice. Such as the Straits of Messina provided a risky, dangerous chokepoint that required a thought-out navigational strategy, the IRTC provided such a strategy to vessels transiting the pirate-averse waters of north Somalia. The IRTC is a key point of reference on the Q6099 Maritime Security Chart for the Red Sea, Gulf of Aden and Arabian Sea (Q6099) that also displays the specially designed High Risk Area (HRA) as well as reference to regional vessel monitoring services and Best Management Practices as performative behavioural requirements for mariners in transit.

The Q6099 was the first internationally available maritime security chart pertaining to the risk of Somali-based piracy and was published on 16 September 2010. Produced by the United Kingdom Hydrographic Office (UKHO), now known as Admiralty UK, the

Q6099 received significant revisions in its 2010 edition, 2015 edition as well as the 2019 edition. The dynamism of the map reflects the political bartering process that designates *where* is considered risky and therefore highlights that the map is a tool with operational consequences in directing navigational behaviours at sea. The Q6099 is one of six specialised security charts that is produced and distributed by Admiralty UK; the others address: the Mediterranean Sea, the Persian Gulf, Karachi to Hong Kong, Singapore to Papua New Guinea and West Africa. The production of maritime security charts is directed by the 'UKHO Security of Navigation, Stabilization, Advice and Training (SONSAT), who are responsible for the collation of content in liaison with external stakeholders including states, marine experts and military personnel' (Anon, UKHO, personal correspondence, 2019) who inform and direct the negotiation process of how risky space is represented in print.

The headquarters of many powerful private maritime industrial actors have remained in the City of London, reflective of a historical legacy of British hegemony at sea. Yet, the power is no longer consolidated within these institutions based on their British affiliation, as they have become privatised and internationalised. For example, Admiralty UK as a trading fund of the UK Ministry of Defence is economically and administratively independent from the UK government despite the UKHO holding a .gov web address. Yet, it is a globally influential chart supplier to vessels used by over 90 per cent of ocean-going vessels (UK Hydrographic Office, 2020). The apparent London-centric maritime monopoly is echoed through the presence of other influential maritime actors within the City of London, such as Lloyd's of London insurance marketplace (that provides the buying and selling of insurance policies for brokers worldwide), the International Group of P&I Clubs (IGP&I), Oil Companies International Marine Forum (OCIMF), the International Chamber of Shipping (ICS), as well as the state wing of organised international maritime governance, the International Maritime Organization (IMO).[1] Yet, in 2018 a study to identify the top maritime nations that focuses on shipping, ports and logistics, maritime finance, law and technology held that China is the top maritime nation, followed by the US. In the report, the UK came eighth (DNV Global/Menon, 2018). The influence of other national interests is evident in the debate that emerged following the release of the Q6099, as the delineation of the HRA and behavioural requirements it provoked proved to be a costly and inconvenient label for the trading states it encompassed.

To determine the HRA, Admiralty UK engages with external shareholders through the Shared Awareness and Deconfliction (SHADE) forum, which has directed and produced much of the counter-piracy literature and decision making in the effort to combat Somali-based piracy. SHADE consists of a broad range of industrial actors, naval missions, peripheral states as well as economically wealthy states who are concerned about the risks their import and exported materials are exposed to in the face of piracy. Both state and industrial actors have engaged with SHADE temporarily, based on the relevance of piracy to their daily business (whether it is the transfer of cargo or seafarer welfare) and politically relevant interests. Egypt, for example, became concerned with the loss of levies from the Suez Canal when vessels had re-routed around the Cape of Africa to avoid the HRA until they realised they lost more money from this through corrupt practices in the Suez itself (Roberts, personal correspondence, 2020). India further argued that no piracy activity was taking place in their territories, and the inclusion of their waters in the HRA was diminishing the reputation of their ports (Bueger, 2018: 629). Where states primarily lobbied alterations to the Q6099 chart through the Contact Group on Piracy off the Coast of Somalia (CGPCS), this lobby was aimed at the large industrial bodies who engaged with the SHADE platform who then inform Admiralty of changes to coordinates that the Q6099 chart displays.

Admiralty UK have defined the most powerful industrial actors in shaping the designation of the HRA as 'the Baltic International Maritime Council (BIMCO), the ICS, INTERCARGO and OCIMF' (Anon UKHO, personal correspondence, 2019). The so-called controversy of the HRA[2] that took place from 2008 onwards revolved around the size of the HRA.

The first representation of the HRA on the Q6099 Maritime Security Chart was released in September 2010 (BMP4, 2011: 94–5). It encompassed the width of the Indian Ocean from East Africa to the Indian sub-continent, inclusive of the territorial waters of several coastal states. The first significant revision to the map was in 2015, where the HRA was reduced and the territory that was previously deemed high risk was replaced by the less intimidating purple hue that depicts the Voluntary Reporting Area (VRA). Transit through this newly constructed spatiality still required seafarers to engage with the monitoring and surveillance structures implemented by the UK Trade Maritime Operations (UKMTO) and the Maritime Security Centre Horn of Africa (MSCHOA) in their transit of the Indian Ocean but without the stigma of *high risk* that was starting to damage the economies of peripheral states. Consequently, it provided a transformative space where seafarers would undertake drills and dress the vessel in preparation for entering the HRA. This provided an important stage in the construction of a temporary deterrence site on board the vessel and in the construction of a field of insecurity.

The most recent (at the time of writing) alteration to the HRA was in May 2019 (UK Hydrographic Office, 2019), where the HRA was reduced and replaced by the VRA once again. This is significant for the Seychelles, which for the first time since before 2008 were declared piracy free, as it reduced the expense of securing a vessel in transit there. The economic impact of both piracy and counter-piracy has been an important factor in the lobby of the CGPCS, based on the association of the HRA as a dangerous, risky space and therefore an expensive space to transit. At the height of successful hijackings in 2011, the maritime industry bore 80.5 per cent of the cost of deterring piracy, at $5.5 billion (Oceans Beyond Piracy, 2011a: 1) for aspects of protection such as re-routing, security equipment, increased speeds, additional insurance requirements, labour and ransoms. States contributed the remaining 19.5 per cent at $1.3 billion (Oceans Beyond Piracy, 2011b) for aspects such as military deployments and prosecutions. The impact of these additional costs were considered in the same way that Aeneas evaluated the impact of re-routing around Sicily against effort and potential losses incurred in being adequately protected in navigating the Straits of Messina. The consequence of many shipowners re-routing and skipping ports was also an economic cost to states who experienced a decline in maritime trade because of their coastline's inclusion in the HRA. Two prime examples include Kenya, where it cost $414 million and Yemen $150 million (Oceans Beyond Piracy, 2011a) per year in trade losses between 2005 and 2011 that is considered a result of their coastlines' inclusion in the HRA.

Therefore, the significance of the Q6099 as a navigational tool has been significant in both the state political and economic impact of piracy as not just a tool that is engaged with by mariners at sea. It is a tool of governance that has encountered significant political debate that has entangled both state and industrial interests in navigating the risk of piracy at sea.

Navigating piracy: operational deterrence practices

The case of the Q6099 shows the extent to which the way the world should be represented to seafarers who navigate the Indian Ocean is a product of shore-based institutional negotiations. As mentioned previously, the Q6099 not only depicts dangerous geographies, but provides detailed instructions of different support structures and protocols for seafarers in transit, one such being reference to the Best Management Practices (BMP) as a practice guide for navigating the HRA. This provides the next stage in the process of insecure spatial production, as the Master then translates this information alongside practice guides to assign specific jobs and temporary alterations to the vessel's architecture.

The focus on the seafarer in performing exceptional navigational practice is important for successful counter piracy, as it is argued that it would take '83 helicopter-equipped vessels to provide an effective one-hour response, when most vessels are only able to request help with a 10 minute warning' (Jones, 2013: 45). As a consequence, crew action has come to play a 'large and statistically significant role in decreasing a ship's relative risk for both robbery and hijacking' (Lewis, 2016: 121) and therefore their behaviours have been identified as having an integral potential in the mitigation of the threat. Written specifically for the seafarer, an analysis of the BMP guides will offer two important contributions to this chapter. First, it will evidence how the BMP has developed as a guide to performing insecure spatial transformation, when used in conjunction with the Q6099 to align behaviour with geography. This will be affirmed by the experiences of the seafarer. Second, it will demonstrate the BMP has experienced several revisions, where the refinement of the threat and the specificity of practical measures is evident in each re-release of guidance and therefore reflects the process of resource reproduction based on information sharing from land to sea. Finally, as drawing on evidence in the most recent edition, the content of BMP5 reflects that as successful piratical attacks have declined, the conglomerate of powerful industries – who had shaped behaviour in the Indian Ocean for more than a decade – are reorienting their perception of danger and risk towards a broader security agenda in pursuit of maintaining the security regime and its structures of governance.

The BMP are a list of practical measures, written as a step-by-step guide for transforming the modern, cargo-oriented design and practices of the modern merchant vessel into a site of piratical deterrence. From the third edition onwards, it is open source and readily available online in a published format with a familiar graphic design that is resonate in the fourth and fifth editions that follow it. The first two editions were, however, criticised in their infancy, for resembling circulars and contextually 'generalizing the modus operandi of pirates and assuming [that] methods of the past' (Hansen, 2012: 566) would be effective. Yet, this deficiency is corrected in the following documents that provide platforms for learning and feedback, which increased the precision of practice in later editions. This highlights that BMP is positioned in the loop of spatial production, bridging land and sea as both a tool for dissemination and feedback that connects the membership of SHADE and the CGPCS with merchant vessels worldwide.

The BMP series emerged from several separate industrial and state efforts to collaborate on advice for mariners who were transiting pirate-averse areas. A representative from the ICS, who had been heavily involved with the drafting process since 2008, explains that the 'core precursor documents for the third, open source BMP was the January 2009 US BMP, the February 2009 MSCHOA BMP as well as the 2018 OCIMF pocket size booklet that was distributed to seafarers' (Anon, ICS, personal correspondence, 2019). Hansen further contributes that IMO Circular 1334 'Advice to Masters' (Hansen, 2012: 566) was a key preceding document containing directive advice and reference to other organisational responses, including the International Maritime Bureau's (IMB) Piracy Reporting Centre (PRC). The range of different advice that was available in the early days of counter piracy reflects that the relationships and collaborations between the industry were not yet consolidated.

The BMP guidance became increasingly refined, streamlined and specific in its prescription of practice because of an increased number of signatories. Table 7.1 shows how between 2009 and 2011, the release of each edition was almost an annual process and the inclusion of new actors has served to enhance the dimensions of deterrence practice to include personal, professional and technical advice. Therefore, as a navigational technology, in each release the BMP entered an increasing number of areas of daily professional and personal life for the seafarer from adjustments to the watch schedule (BMP1, 2009), to behavioural advice in the event of a hijacking as well as navigational behaviours such as specialised manoeuvres and specific vessel speeds (BMP3, 2010; BMP4, 2011; BMP5, 2018). The increase in the number of signatories and the impact of their contribution reflects that BMP is a familiar yet living document, which in each release further infiltrates insecure practice into the daily routine of the seafarer.

BMP1 was considered non-specific and lacking in its understanding of the dynamism of the piratical threat. The document proposed the Master sailed at increased speeds, but didn't state what these should be (Hansen, 2012: 565) and referred to the employment of 'additional resources' (BMP1, 2009) as assumed and unspecified. BMP2, released in August 2009, was a product of greater industry collaboration, seafarer feedback and naval surveillance through information centres such as MSCHOA and SHADE. This is evident in the inclusion of more specific advice, such as 'adhering to designated transit corridors, installing remote-controlled fire-hoses, and darkening the ship' (BMP2 as cited in Hansen, 2012: 566). Yet, BMP2 still resembled a circular as opposed to a bespoke handbook. BMP3, released in 2010, advanced this specificity, drawing on the experience, knowledge and advice of an increased number of industry-affected partners who were experiencing losses to piracy upon its release in 2010. Very specific practices such as 'travelling at eighteen knots … employing the use of bulletproof glass as well as installing barriers and razor wire' (Hansen, 2012: 566) around the deck of the vessel are indicative of the production and reproduction of navigational resources that produce insecure spaces,

Table 7.1 Cumulative list of BMP signatory organisations 2009–2018

BMP1 (February 2009)	BMP2 (August 2009)	BMP3 (August 2010)	BMP4 (August 2011)	BMP5 (June 2018)
1. Intertanko	BMP1 Signatories including:	BMP2 Signatories including:	BMP3 Signatories including:	BMP4 Signatories including:
2. International Chamber of Shipping (ICS)	12. International Transport Workers' Federation (ITF)	16. International Shipping Federation (ISF)	22. International Maritime Employers' Council (IMEC)	27. Chemical Distribution Initiative (CDI)
3. Oil Companies International Marine Forum (OCIMF)	13. Maritime Security Centre Horn of Africa (MSCHOA)	17. International Parcel Tanker Association (IPTA)	23. InterManager	28. International Federation of Shipmasters' Associations (IFSMA)
4. Baltic International Maritime Council (BIMCO)	14. UK Maritime Trade Operations (UKMTO)	18. European Union Naval Force (EUNAVFOR)	24. Mission to Seafarers	29. International Marine Contractors Association (IMCA)
5. Society of International Gas Tanker and Terminal Operators (SIGTTO)	15. Maritime Liaison Office (MARLO)	19. Combined Maritime Forces (CMF)	25. World Shipping Council (WSC)	30. International Maritime Employers Council Ltd (IMEC)
6. Intercargo		20. North Atlantic Treaty Organization (NATO) Operation Ocean Shield	26. Interpol	31. The Sailors Society
7. International Group of P&I Clubs (IGP&I)		21. NATO Shipping Centre		32. Contact Group on Piracy off the Coast of Somalia (CGPCS)
8. Cruise Lines International Association (CLIA)				33. Naval Co-operation and Guidance for Shipping (NCAGS)
9. International Union of Marine Insurance (IUMI)				
10. Joint War Committee (JWC)				
11. International Maritime Bureau (IMB)				

through the narrowing down of specific practices based on shore-based expertise, collaboration and vessel-based feedback.

The release of BMP4 in August 2011 marks a critical moment in the BMP series, as it became the core referenced document for seven years, until the release of BMP5 in 2018. The International Maritime Employers Council (IMEC), InterManager and the Mission to Seafarers joined as additional signatories to BMP4 (see Table 7.1) and the seafarer safety perspective is therefore a resonate theme throughout the document. Behavioural advice for seafarers to follow in the event of an attack is presented in simple English and displayed in a bold and colourful fashion. Entire pages, such as the 'Aide Memoire – Avoid Being a Victim of Piracy' (BMP4, 2011: vi), are designed to be read and understood by seafarers of all nationalities and abilities. The distribution of BMP4 was no longer targeted at those sat in offices ashore, or for the quick glance of the Master when checking their emails. But its production in a pocket-size format (drawing on OCIMF's earlier guidance) reflects the ambitions of the authors for BMP to have a place in the pocket 'of every mariner in the world' (Bueger, 2018: 626). By releasing BMP to sea, not only did mariners have the Q6099 as a frame of reference as to where piracy was likely to take place, but a detailed list of instructions in preparing themselves and the vessel for a potential encounter and, as such, the architectural direction to construct a temporary and mobile insecure spatiality.

The specific authorship of the BMP series has been considered an ambiguous and opaque process, yet the release of the fifth edition and personal accounts given by senior representatives from the IGP&I Clubs suggest otherwise. Bueger's work on the CGPC's engagement with the maritime industry concludes that much of the work relating to the drafting of BMP is outside of the public domain in a 'London based black box' (Bueger, 2018: 625) with little detail as to who actually agreed on the final wording of the documents. While previous editions failed to state the authorship of the document, instead referring only to its signatories, the fifth edition includes the named authors as 'BIMCO, ICS, IGP&I Clubs, INTERTANKO and OCIMF' (BMP5, 2018). Two senior representatives from the IGP&I Clubs, Simon and Harry, both contributed to the drafting of the BMP series and have worked on every edition that has been released.[3]

The IGP&I Club is a mutual insurance group, with a membership of thirteen mutual insurance clubs. The IGP&I has an outreach to over 90 per cent of ocean-going tonnage (Harry, personal correspondence, 2018) and despite not considering themselves a 'big cog in the wheel' (Harry, personal correspondence, 2018) their value to the production and reproduction of insecure space comes from their dissemination and consultation capabilities. Harry and Simon assert that much of the operational processes that were included in the BMP drafting process were deferred to the shipowner associations (such as OCIMF and BIMCO), but they commented on aspects such as the safety of the process from an insurance perspective (Simon, personal correspondence, 2018): 'Through our committees or through us, we have a chairman from the biggest Clubs, they attend meetings with other associations, feedback with what has been discussed and then we feed those back to the others [authors], with the recommendations they have suggested' (Harry, personal correspondence, 2018).

By calling on their wider membership of shipowners, the IGP&I Club was able to draw on required information from different vessels while in transit, as well as access live and emerging matters from seafarers directly. This posits their role as a significant nodal point of the cyclical production and reproduction of insecure space, as they capture information, translate it in relation to broader industry interests, and then contribute to how it is incorporated into documents such as BMP, which are then communicated back to sea.

The value of their input to BMP was therefore not exclusively based on their dissemination capabilities, but their capacity to learn from previous experiences as well as suggest new directions for the security regime to reorient. In doing so, the networks remained relevant to those who valued their information-sharing capacities and have built businesses and livelihoods that efforts to deter piracy have created (Roberts, personal correspondence, 2020). Simon argues that the information sharing structures have accounted for there 'being a better picture now, [where] BMP was originally geared up towards piracy, it's now about other things and will continue to evolve to address things potentially like cyber risks and possible links to migration at sea' (Simon, personal correspondence, 2018). The development of a website and the ambiguity around what might be addressed in the future suggests that although future potential threats are not yet specifically defined, there is an appetite and consequently structures in place to identify *what's next* for the same networks that have directed counter-piracy governance.[4]

BMP5 and the experience of the modern mariner

Released in June 2018, BMP5 has been circulated to mariners who are traversing waters six years following the peak of successful piratical hijackings. The endeavour to construct a new, yet ambiguous, security landscape resonates throughout the document, and is in line with Harry's reflection on the BMP's reorientation towards new security concerns. The reorientation towards security beyond piracy is apparent through a discursive shift in the document. The first aspect of BMP5 that reflects this is the extended title that claims the purpose of the document is to 'Deter Piracy and Enhance Maritime Security in the Gulf of Aden, Indian Ocean and Arabian Sea' (BMP5, 2018). Previous editions have centred their titles exclusively around the piratical threat; the shift to generally enhancing maritime security alludes to the openness of the document in its focus. This is a sentiment echoed by Simon, who admits there is no 'single list as to what could be addressed in the future' (Simon, personal correspondence, 2018). Continuing the shift away from a piratical focus, potential adversaries are referred to 'attackers' (BMP5, 2018) as opposed to 'suspected hijackers' or 'pirates', broadening the remit as to *who* is considered a credible threat to navigation and transit. In providing a platform for capturing new paths of insecurity, new annexes C and D offer the seafarer the opportunity to report 'suspicious activity' (BMP5, 2018: 50) such as sightings of a 'vessel towing multiple skiffs', 'Not flying a flag' or with 'packages hanging outboard of a vessel' (BMP5, 2018: 36). Although the irregular presence of a skiff or of flags may be accredited to the piratical threat, hanging packages are assigned more so with water-borne improvised explosive devices (WBIEDs) that are associated with both Iranian-backed Houthi rebels (SOLACE Global, 2018) and Iran.

The experiences of individual seafarers offer rich accounts of navigating the HRA and transforming merchant vessels into spaces of insecurity, through engaging with the Q6099 and BMP series. Their contribution is threefold; they provide evidence as to how the navigational resources are translated and performed at sea, how they provide feedback to the shore-based security actors as well as provide evidence that the Houthi and Iranian threat to merchant shipping is now framing risk as opposed to exclusively piracy.

Each vessel is commanded by the Master as the overriding figure of authority, so the translation and designation of BMP responsibilities differs on each vessel based on personality, interpretation of the threat, the requirements of the vessel as well as the company capacity to afford different aspects of protection. Captain Ranjish Joshi, an Indian national with over twenty years experience at sea, argues that his approach to translating counter-piracy advice is to gather all advice he is given (Joshi, personal correspondence, 2019). This may include company advice, vessel-specific advice as well as the BMP. He then draws on the operational requirements of the most stringent advice to draw up his ship's security plan (SSP). The benefit is that by following the strictest advice, he is covered in all circumstances and his crew is offered the best protection. In feeding back to his company, he explains that they 'run competitions in the company magazine and offer incentives' (Joshi, personal correspondence, 2019) to suggest innovative ideas and novel practices to enhance security. Both his approach to writing the SSP as well as company feedback corresponds to the process as cyclical, whereby there is a familiar stream of communication that connects vessel performance and the shore-based construction of navigational technologies.

Captain Andrew Oddy takes a similar approach in gathering all advice that is available to him, yet his emphasis is more on ensuring that each member of crew is aware of their responsibilities. This account reflects that although BMP and Q6099 are familiar technologies for merchant mariners, the unpredictability of human behaviour and personalities of those with authority means that translation and performance is not a uniform or universal experience. Captain Oddy draws on his childhood growing up in Cyprus and South Africa during times of conflict as his rationale for prioritising that his crew have a calm and informed approach to their assigned tasks, drawn from BMP guidance. He argues that order and routine is the core element of its success, so that no one 'loses their heads and ends up shot just to make them shut up' (Oddy, personal correspondence, 2019). This is a sentiment shared by Cadet Connel,[5] who transited the HRA in mid-2018. His training for engagement with BMP was administered by privately contracted armed security personnel (PCASP) and as a cadet in training he felt that his lack of experience meant that being given clear instructions was an important part of making sure tasks were performed efficiently. This included several drills, meetings as well as scare videos that depicted scenarios of what could go wrong if BMP is not followed diligently. Similarly to Captain Joshi, Cadet Connel was offered the opportunity on several occasions to feed back following group tasks such as setting up the safe space (citadel) as well as dressing up the ship with hoses and razor wire. He explains that a fellow cadet suggested 'greasing the poles' (Connel, personal correspondence, 2019) around the vessel to stop anyone climbing them and that this was implemented the following day. This is another example as to how constructing insecurity at sea has been a reflective and dynamic process both on land and at sea.

In addressing emerging threats, Madeleine Wolczko is a Second Officer who experienced the HRA in late 2018. It was her responsibility to plan the vessel's passage as a key component of navigation. She argues that the risks she was considering in her passage plan were not exclusive to the piratical threat defined by the HRA, but by the presence of the Houthi Rebels in the Bab-Al-Mandeb: 'they had put out a letter that said anything that comes within 40nm of Al-Hudaydah … will be fired upon. I had to re-do the track line. At that time, the rebels were more our concern than the pirates were' (Wolczko, personal correspondence, 2019).

The Houthi rebels pose a live and credible threat to merchant ships in the Bab-Al-Mandeb, while the Iranian National Guard poses a threat through their attempts to control access (SOLACE Global, 2018) to the Straits of Hormuz. Officer Wolczko's experience saw that her fellow crew engaged with the technologies that had been initially designed for addressing piracy as a means of mitigating this new threat. Although counter-piracy practices may not be appropriate for mitigating new threats, they are familiar and provide the seafarer with a sense of routine that eases the apprehension that a new and dangerous threat poses, while simultaneously through their use against new threats carving out new spaces for insecurity to encompass. This included the employment of PCASPs as well as displaying 'sticky tape' (Wolczko, personal correspondence, 2019) on the bridge windows to protect the crew from shattering glass if a WBEID or rocket was experienced. In applying the framework of insecure spatial production as a loop of communication and performance, the post-piratical future looks likely to include these threats as a key focus for the traditionally piracy-focused security actors.

Conclusion

We navigate the seas by drawing on the familiarity of our landed politics that are embedded within maps and navigational tools. In our transit, we take these politics to sea and they underpin our designation of risk and navigation of security while working with the sea to fulfil our transit needs. As such, navigation is a political engagement. This chapter has suggested that a Paris School conceptualisation of insecurity combined with a dynamic and cyclical process of spatialisation can help explain how spaces of insecurity are constructed aboard the mobile vessel in its propulsion on the materially liquid sea. Thinking of this process as dynamic removes the imagination of the sea as time *wasted*, thinking of it instead as time *spent*. This includes a transformative process where mariners engage with maps and guidance to navigate risky spaces. Each have different areas of expertise that have, alongside shore-based interests, been useful in the capture of pirates and in the development of new guidance for mitigating piracy. In the contemporary era, protecting global shipping means securing the backbone of the global economy.

Analysing maps and navigational guidance from the Babylonian and polytheistic ancient periods reflect how landed political philosophies and familiar frames of reference are embedded within them as tools. The first 'map of the world' delineates security with familiarity, where Babylon is designated as safe, known and experienced. Yet in travelling further away, insecurity is framed by the unknown. In the modern counter-piratical context, designing a HRA on a familiar navigational chart has facilitated the production of insecure space by delineating a spatial frame of reference for exceptional practices to take place as well as delineating *where* the vessel should undergo a social and physical transformation.

The contribution of this chapter in understanding how we tame the sea within our innately landed IR imaginations is not just a balance of the precariousness of keeping a vessel afloat and mobile within its liquidity. Nor is it an attempt at throwing a generalised and assumptive anchor of IR theory to sea. It is a recognition that shore-based political debates and regimes have informed the construction of maps and practice guides that have operational consequences in the construction of insecure spaces and the navigation of risk. Yet, the sea's liquidity provides the environment for this process to be mobile, atemporal and reliant on the transformative performance of the seafarer. This is a process that can be revised and further observed by drawing on navigational technologies, their construction and performance.

Notes

1 While the headquarters of the United Nations is New York City, NY, USA, the IMO has been situated in London since its conception in 1948.
2 The politics of which are presented in detail in Bueger (2018).
3 Pseudonyms have been assigned to preserve anonymity.
4 Maritimeglobalsecurity.org is a website that has been set up by BIMCO, IGP&I, Intertanko, OCIMF, InterManager, Intercargo and the ICS.
5 Pseudonym.

References

Anderson, J., and K. Peters (2014). *Water Worlds: Human Geographies of the Ocean*, Farnham: Ashgate.

Andrews, E. (2016). '8 remarkable early maps', www.history.com/news/8-remarkable-early-maps (accessed 10 February 2022).

Apollonius of Rhodes (1998). *The Argonautica*, R. Hunter, ed., Oxford: Oxford University Press.

Bigo, D. (2016). 'Rethinking security at the crossroad of international relations and criminology', *British Journal of Criminology* 56(6): 1068–86.

Bigo, D., and A. Tsoukala (2008). *Terror, Insecurity and Liberty: Illiberal Practices of Liberal Regimes after 9/11*, Abingdon: Routledge.

Blomberg, S. B., R. Fernholz and C. Levin (2013). 'Terrorism and the invisible hook', *Southern Economic Journal* 79(4): 849–63.

BMP1 (2009). *Best Management Practices to Deter Piracy in the Gulf of Aden and off the Coast of Somalia*, unpublished.

BMP2 (2009). *Best Management Practices to Deter Piracy in the Gulf of Aden and off the Coast of Somalia*, 2nd edn, unpublished.

BMP3 (2010). *Best Management Practices to Deter Piracy off the Coast of Somalia and in the Arabian Sea Area*, 3rd edn, Edinburgh: Witherby Publishing Group.

BMP4 (2011). *Best Management Practices for Protection against Somalia Based Piracy*, 4th edn, Edinburgh: Witherby Publishing Group.

BMP5 (2018). *BMP5 – Best Management Practices to Deter Piracy and Enhance Maritime Security in the Red Sea, Gulf of Aden, Indian Ocean and Arabian Sea*, 5th edn, Edinburgh: Witherby Publishing Group.

Boren, J. (2014). 'Negligent prosecution: why pirates are wreaking havoc on international trade and how to stop it', *European Journal of Law Reform* 16(9): 19–40.

Bourne, M. (2013). *Understanding Security*, Basingstoke: Palgrave Macmillan.

Bueger, C. (2013). 'Practice, pirates and coast guards: the grand narrative of Somali piracy', *Third World Quarterly* 34(10): 1811–27.

Bueger, C. (2018). 'Territory, authority, expertise: global governance and the counter-piracy assemblage', *European Journal of International Relations* 24(3): 614–37.

Connel, R. W. (2019). Interview with J. K. Simonds, 8 January 2019.

Cummins, J. D., and P. Trainar (2009). 'Securitization, insurance, and reinsurance', *Journal of Risk and Insurance* 76(3): 463–92.

Dalley, S. (2000). *Myths from Mesopotamia Creation, the Flood, Gilgamesh, and Others*, 2nd edn, Oxford: Oxford University Press.

Danzer, G. A. (1991). 'Part 2: the earliest world map, Babylonia c. 500 B.C.', *Social Education* 55(1): 304–5.

DNV Global/Menon (2018). *Top Maritime Nations of the World 2018*, Oslo: DNV Global/Menon, www.menon.no/wp-content/uploads/2018-84-LMN-2018.pdf (accessed 10 February 2022).

Flick, U. (2018). *An Introduction to Qualitative Research*, 6th edn, Thousand Oaks, CA: Sage.

Germond, B. (2015). 'The geopolitical dimension of maritime security', *Marine Policy* 54, 137–42.

Gilmer, B. (2019). 'Invisible pirates: women and the gendered roles of Somali piracy', *Feminist Criminology* 14(3): 371–88.

Hansen, S. J. (2012). 'The evolution of best management practices in the civil maritime sector', *Studies in Conflict and Terrorism* 35(7–8), 562–9.

Harry (2018). Interview with J. K. Simonds (28 September 2018).

Homer (2002). *The Odyssey*, 2nd edn, R. Fagles, ed., London: Penguin Books.

Huysmans, J. (2006). *The Politics of Insecurity: Fear, Migration and Asylum in the EU*, B. Buzan and R. Little, eds, London: Routledge.

Huysmans, J., A. Dobson and R. Prokhovnik, eds (2006). *The Politics of Protection: Sites of Insecurity and Political Agency*, 1st edn, Oxon: Routledge.

Ivancovich, J. S. (2015). 'Pirate trails: tracking the illicit financial flows from pirate activities off the Horn of Africa', *The European Journal of Development Research* 27(3), 476–8, http://documents.worldbank.org/curated/en/408451468010486316/pdf/812320PUB0REVI00Box37983 8B00PUBLIC0.pdf (accessed 14 June 2021).

Jones, S. (2013). *Maritime Security Handbook: Coping with Piracy*, 1st edn, London: The Nautical Institute.

Joshi, R. (2019). Interview with J. K. Simonds, (4 February 2019).

Keenan, J. (2014). 'Puntland is for pirates', *Foreign Policy* 205: 1.

Lansing, P., and Petersen, M. (2011). 'Ship-owners and the twenty-first century Somali pirate: the business ethics of ransom payment', *Journal of Business Ethics* 102(3): 507–16.

Lefebvre, H. (1974). *The Production of Space*, 3rd edn, Oxford: Blackwell.

Lefebvre, H. (2004). *Rhythmanalysis: Space, Time and Everyday Life*, 3rd edn, S. Elden, ed., London: Bloomsbury.

Lewis, J. S. (2016). 'Maritime piracy confrontations across the globe: can crew action shape the outcomes?' *Marine Policy* 64, 116–22.

Levinson, M. (2006). *The Box: How the Shipping Container Made the World Smaller and the World Economy Bigger*, 1st edn, Princeton, NJ: Princeton University Press.

Linebaugh, P., and Rediker, M. (2012). *The Many-headed Hydra: The Hidden History of the Revolutionary atlantic*, London: Verso.

Netz, R. (2004). *Barbed Wire: An Ecology of Modernity*, 1st edn, Middletown, CT: Wesleyan.

De Nevers, R. (2015). 'Sovereignty at sea: states and security in the maritime domain', *Security Studies* 24(4): 597–630.

Oceans Beyond Piracy (2011a). 'The economic cost of piracy 2010', The One Earth Foundation, https://oneearthfuture.org/research-analysis/economic-cost-maritime-piracy-2010 (accessed 10 February 2022).

Oceans Beyond Piracy (2011b). *The State of Maritime Piracy 2011: Assessing the Economic and Human Cost*. Denver, CO: One Earth Future Foundation.

Oddy, A (2019). Interview with J. K. Simonds (23 May 2019).

Patton, P. (2013). 'From resistance to government: Foucault's lectures 1976–1979', in C. Falzon, T. O'Leary and J. Sawicki, eds, *A Companion to Foucault*, 1st edn, Chichester: Wiley-Blackwell, 172–88.

Roberts, N. (2020). Interview with J. K. Simonds (2 April 2020).

Salter, M. (2008). *Politics at the Airport*, http://onlinelibrary.wiley.com/doi/10.1002/cbdv.200490137/abstract (accessed 14 June 2021).

Simon (2018). Interview with J. K. Simonds (28 September 2018).

SOLACE Global (2018). *Water Borne Improvised Explosive Devices (WBIED)*, www.solaceglobal.com/wp-content/uploads/2018/03/20180308-Maritime-Advisory-History-and-Usage-of-Water-Borne-Improvised-Explosive-Devices-1.pdf (accessed 20 March 2020).

Steinberg, P. E. (2001). *The Social Construction of the Ocean*, 1st edn, New York: Cambridge Studies in International Relations.

Steinberg, P., and Peters, K. (2015). 'Wet ontologies, fluid spaces: giving depth to volume through oceanic thinking', *Environment and Planning D: Society and Space* 33(2): 247–64.

UK Hydrographic Office (2019). *Maritime Security Chart: Red Sea, Gulf of Aden and Arabian Sea*, https://on-shore.mschoa.org/media/1413/hra.jpg (accessed 7 July 2020).

UK Hydrographic Office (2020). 'Admiralty Maritime Data Solutions', www.admiralty.co.uk/ukho/About-Us (accessed 17 May 2020).

Van Munster, R. (2007). 'Review essay: security on a shoestring: a hitchhiker's guide to critical schools of security in Europe', *Cooperation and Conflict*, 42(2): 235–43.

Verderame, L. (2020). 'The sea in Sumerian literature', *Water History* 12(1): 75–91, https://doi.org/10.1007/s12685-020-00244-6 (accessed 14 June 2021).

Virgil (2006). *The Aeneid*, 1st edn, R. Fagles and B. Knox, eds, London: Penguin Books.

Wæver, O. (2011). 'Politics, security, theory', *Security Dialogue*, 42 (4–5): 465–80.

Walt, S. M. (1998). 'International relations: one world, many theories', *Foreign Policy* 110: 29–46.

Wolczko, M. (2019). Interview with J. K. Simonds (9 March 2019).

Wood, D. (1992). *The Power of Maps*, 1st edn, New York: The Guildford Press.

Woods, C. (2012). Sons of the sun: the mythological foundations of the first dynasty of Uruk, *Journal of Ancient Near Eastern Religions* 12(1): 78–96.

8

Obligations *erga omnes* and the common heritage of mankind under the Law of the Sea Convention

Filippa Sofia Braarud

In the latter half of the twentieth century, the term 'common heritage of mankind' was coined as a promise to the international community, stipulating that all states would benefit equally from the areas that fell within its scope. The regime grew to encompass the deep seabed among other areas that lay beyond the sovereign jurisdiction of any one state. The deep seabed, also known as the Area, has long been left relatively idle due to its inaccessibility and the difficulties in making meaningful use of it. Today, however, technological advances have made industrial deep-sea mining operations possible, and the increasing demand for minerals and metals for the development of new technologies while land-based mines are facing depletion has led to a surging interest in deep-sea mining. The ecological consequences of industrial mining operations to the marine ecosystems on the deep seabed remain understudied, and these potential ecological risks have prompted a search for new legal mechanisms for states to seek redress should the Area be unlawfully polluted, appropriated or exploited. Although still a relatively ambiguous concept in international law, obligations *erga omnes* have emerged in cases where the norm breached would be redundant if no state could claim to have a legal interest in defending it. This chapter will analyse the legal grounds underpinning the presumption that the common heritage designation of the deep seabed gives rise to *erga omnes* obligations as technology is slowly making it accessible for exploitation. This chapter addresses the important role the sea plays in international politics, and the third trope regarding the extraction and control of maritime resources in particular, by addressing how technologically advanced countries are set to dominate this newly accessible wealth and how this can be controlled through international law. Moreover, it highlights the tendency in global politics of treating the sea conceptually more as land in the traditional sense of sovereignty, moving beyond Hugo Grotius' idea of *Mare Liberum* and the perception of the deep seabed as *terra nullius*. Ultimately, the domination of the resources on the deep seabed may well be what will define power politics and international relations in the next century to come.

Over the past few decades, international law has entrenched the protection of the deep seabed as part of the Common Heritage of Mankind (CHM) in the United Nations Convention on the Law of the Sea (UNCLOS) Article 1(1) as 'the Area', composed by the deep seabed beyond the national jurisdiction of any country (UNCLOS 1982, Art. 1(1)). Deep-sea mining involves mining rich deposits of minerals on this deep seabed under the high seas and is slowly emerging as a viable industry. As there has been little mining activity in the Area to date, the environmental risks involved with mining operations have been scarcely studied and more research is now called for as industrial mining operations are becoming a reality. This chapter will focus on the legal challenges of protecting the Area as it is designated as common heritage under the UNCLOS. One main challenge is that the size of the Area is negatively defined by the delimitation of the continental shelf, as what is 'left over' after all littoral states have claimed their continental shelf. The question this chapter attempts to answer is whether, by designating the Area as common heritage of mankind, the UNCLOS confers upon all State Parties an obligation towards all State Parties – an obligation *erga omnes* – to respect this designation. If it does, then how and by whom can these obligations be enforced should a State party exploit the Area in an unlawful manner? Finally, to whom would the reparations sought be allocated in case a state is found to be liable, given that all states enjoy an equal right to the Area and its resources?

As the traditional bilateral structure of international law is becoming increasingly challenged by attempts to develop a legal order based on community interest, international legal scholars are trying to establish the scope and application of obligations *erga omnes* and explore their potential for allowing third-party intervention in case of unlawful conduct. Since the very designation of an area as CHM implies that its protection is in the interest of all states, what follows is that all states ought to have the right to take legal action against unlawful appropriation or harm. This would be in line with the well-established doctrine *ubi jus ibi remedium* – every right necessitates a remedy. This chapter analyses to what extent this reasoning holds true when applied to the Area, and how much support such a proposition currently has in the existing literature. Due to the historical primacy of states in public international law, this chapter will focus on state responsibility primarily and not consider the responsibility of international organisations and private companies, as the law on responsibility of private companies and organisations is still under development. First, I introduce the theory, history and current status of the common heritage of mankind. From there, I move on to invoking the doctrine of *erga omnes*, its legal grounds and current debates as to its applicability. Third, I assess the avenues for jurisdiction, state responsibility, legal standing and remedies, before finally presenting a summary and analysis of the foregoing parts.

The origins of the common heritage of mankind

The term 'common heritage of mankind' remains at times ambiguous and intangible. In international law, it is used to describe the parts of the global commons that are areas beyond national jurisdiction, that are not subject to national appropriation and designated to be used in such a manner that their resources fall within the interests of both current and future generations. The philosophical reasoning behind this term stems from Hugo Grotius' famous 1608 *Mare Liberum*, in which he proposed that the sea is 'the common property of all', and that the seas should be kept free from appropriation so that all states could have an equal right to access it (Grotius, 1916 [1608]: 28). The term 'common heritage of mankind', as we know it today, however, owes its initial recognition in part to a speech by the Maltese Ambassador Pardo in November 1967, as he addressed the First Committee of the UN General Assembly and proposed that the seabed and ocean floor should be designated as 'common heritage of mankind' (Egede, 2014: 1–2). This speech echoed the general concerns of his time to protect these areas from the 'tragedy of the commons', coined by Garrett Hardin in his 1968 paper with the same name (Hardin, 1968). This was the impetus that led to the Third UN Conference on the Law of the Sea in 1970, during which the Declaration of Principles Governing the Seabed and Ocean Floor was signed, establishing the deep seabed as common heritage to be used only for peaceful purposes (UN General Assembly Res. 2749). This was ultimately included in the UNCLOS of 1982, in which Article 136 states that 'the Area and its resources are the common heritage of mankind' (UNCLOS 1982, Art. 136). The jurisdiction over these resources, defined as 'solid, liquid or gaseous resources, *in situ* in the Area at or beneath the seabed' are thus granted 'to mankind as a whole'. The classification of the Area as CHM followed a trend within international law at the time to seek a more community-centred legal regime as opposed to the traditional, bilateral structure of international law. It also challenged the traditional concept of sovereignty as it conferred this territorial right universally upon all states alike, which ultimately resulted in an ambiguous legal standing in international law.

That the UNCLOS placed the resources located on the deep seabed and ocean floor beyond national appropriation was a historical development in international law that can best be understood in the context of its time. The development of the concept of the common heritage of mankind was to a large extent a result of the historical tensions between developed and developing states at the discussions of the various UN Conferences of the Law of the Sea. As an illustration thereof, the discussion surrounding the introduction of this proposal gives an important insight into the political reality and zeitgeist preceding the enactment of the UNCLOS, including the CHM notion as it stands today. At the time, an increasing number of developing states namely sought to influence the decision-making processes with regard to the law of the sea, to prevent a 'scramble of the seabed' by favouring community norms. Many of these developing states had just untangled themselves of the colonial ties of Western imperialists and opposed placing individual economic security above community security as they experienced this as a threat of the dominating interest of the free market economy of industrialised states (Pinto, 1983: 230). Especially the 'Group of 77', a loose coalition of developing states including important littoral states like the Philippines, Tunisia,

Indonesia, Cuba and Panama, to mention but a few, sought to form a new law of the sea in favour of the international community as a whole (Kasa et al., 2008: 8). The concerns of the Group of 77 were for example showcased in the informal attempt by Chile to include the principle of CHM as a *jus cogens* norm; norms so important that no derogation to them is permitted:

> The states parties to the present Convention accept and recognize on behalf of the international community as a whole that the provision relating to the common heritage of mankind set out in Art. 136 is a *peremptory norm* of general international law *from which no derogation is permitted* and which, consequently, can be modified only by a subsequent norm of general international law having the same character. (UN Doc. A/Conf.62, GP 9, 5 August 1980; Platzoder, 1987: 302; emphasis added)

This statement, pushing for the title of CHM to fall under the *jus cogens* category, alongside the prohibition of genocide, crimes against humanity and war crimes, demonstrates the importance these states attributed to the CHM at the time. Although many states supported this proposal, a consensus proved impossible to reach at the time due to the opposition of a small number of Western states (UN Doc. A/Conf. 62/L. 58, XIV UNCLOS, 129). Yet this statement carried great significance in and of itself, and ultimately became reformulated in UNCLOS Article 311(6) as it currently reads:

> Parties agree that there shall be no amendments to the basic principle relating to the common heritage of mankind set forth in article 136 and that they shall not be party to any agreement in derogation thereof. (UNCLOS 1982, Art 311(6))

Understanding this historical struggle for placing the Area beyond national appropriation is important when grasping certain states' continued scepticism with respect to the workings of the International Seabed Authority, the actions of industrial states and, ultimately, the reasons why enjoying *erga omnes* standing would be a powerful tool should they find the exploitation of the deep seabed to be conducted unlawfully.

In order to make sure that the resources located on the deep seabed are managed responsibly and equitably, Article 156 of the UNCLOS establishes and grants jurisdiction over the Area to the ISA (UNCLOS 1982, Art. 156). The ISA's mandate also includes the promotion of research and environmental protection of the resources in the Area, as well as managing a framework for knowledge sharing between developed and developing states. Although some scholars have found the current ISA to be cost-effective, with well-defined powers and a stable composition of the Council and Finance Committee, others note that the 1994 Implementation Agreement established a 'dysfunctional' ISA altogether (Scholtz, 2008: 277). The fact that the Council's voting system grants veto rights to three industrialised states, and that the Financial Committee, dominated by developed states, can use 'cost-efficiency' as a reason to suspend the governing bodies' sessions, have led to the ISA being criticised for not living up to its communal aspirations (Borgese, 1995: 45; Pinto, 1996: 249–68). Among the most vocal opponents to the 1994 Implementation Agreement establishing the ISA, legal scholar Anand noted that:

> The international community bent backwards to make sure that Washington was satisfied ... Although the area of the deep seabed beyond the limits of national jurisdiction is still called and declared as the common heritage of mankind, the term has lost its original meaning and substance when it symbolized the interests, needs, hopes and aspirations of a large number of poor peoples. The deep seabed will now be exploited on commercial terms, irrespective of the needs and interests of the weak members of the international community. (Anand, 1997: 16–18)

This statement reflects the disillusionment of many states who initially attempted to include the common heritage principle in the New International Economic Order to remedy their colonial past (Tuerk, 2012: 41). Not only does scepticism still prevail with respect to the ISA's lack of transparency and democratic standards, there is also uncertainty as to the actual scope of the Area and its authority on this matter. As ICJ Judge Gaja points out, nowhere in the UNCLOS is the right to decide upon the scope of the Area conferred to the ISA, and thus only states may invoke other states for their alleged unlawfulness should they make claims beyond the 200-nautical-mile limit (Gaja, 2014: 174). As such, it seems that no state or entity would be able to claim legal standing should a littoral state unilaterally and illegitimately expand their continental shelf and infringe on the Area (Wolfrum, 2011: 1141). This raises important questions as to whether and how states may seek to protect the Area, either based on individual or collective interest.

Until recently, the deep seabed has been left relatively idle due to lack of technology and investments in projects to explore and exploit its resources. However, there is currently a surging global interest in exploring the deep seabed due to the depletion of land-based mineral deposits as well as promising estimates regarding the value of the rare minerals located in the Area (Cruickshank, 2016). To date, the ISA has issued thirty-one contracts for the exploration of deep-sea mineral resources in the Area, related to mining of manganese nodules, massive sulphides and ferromanganese crusts, all sedentary formations that are rich in mineral content (ISA Exploration Contracts, 2022). The issue of the ISA has been to provide investment security for investors to deploy large-scale projects under its supervision, as investors have seemed to find the lack of clarity and predictability regarding the process to obtain exclusive rights to mine in the Area an obstacle for large-scale investment (Young, 1989: 129). Paradoxically, if the common heritage of mankind should benefit anyone in order to benefit all, these uncertainties need to be addressed in the legal framework governing the Area, as the revenues of commercial mining are to be shared equitably between all states. The issue thus remains to make sure that appropriate safeguards are in place, to grant security to potential investors and all state beneficiaries alike. The most pressing concerns regarding the protection of the Area are issues pertaining to the delimitation of the outer continental shelf, and deep-sea mining operations that would fail to comply with established environmental standards or the ISA's Mining Code. To demonstrate the urgency of clarifying this legal regime, on 25 June 2021, the president of Nauru requested the ISA to 'complete the adoption of rules, regulations and procedures required to facilitate the approval of plans of work for exploitation in the area within two years', meaning that the ISA will have to clarify these by the end of June 2023 (ISA Press Release, 2021).

Protection of the area through obligations *erga omnes*

Based on the designation of the Area as CHM, and the fact that all states in theory can be said to have a vested interest in making sure that no breach of such obligations to preserve it occurs, this chapter explores whether it follows that such obligations should be recognised as obligations *erga omnes* under international law.

The Latin doctrine '*erga omnes*' literally translates as 'in relation to everyone' and refers to laws that are considered obligations that states have towards all states in the international community. In public international law, *erga omnes* norms differ from the traditional bilateral structure that governs the relationship between states in the sense that all states can engage in countermeasures against a state that violates the obligation imposed by such norms. In an ordinary, bilateral structure, only a state that can prove to have been directly harmed will enjoy legal standing in order to commence procedures (Posner, 2008: 2). In cases where obligations *erga omnes* are invoked, however, all states can in theory claim legal standing.

Traditionally, international law was exclusively bilateral in the sense that rights and obligations principally arise between two states. This strictly bilateral structure was however altered in the 1969 Vienna Convention on the Law of Treaties, where Article 53 recognised a set of peremptory norms – *jus cogens* norms – to which no derogation could be accepted (The Vienna Convention 1969, Art. 53). This marked the beginning of a conceptual change towards a more community-centred trend within international law. The concepts of *erga omnes* norms and *jus cogens* norms are interlinked in many ways; however, they should not be seen as interchangeable. The first time *erga omnes* obligations were invoked by the ICJ was in the *Barcelona Traction, Light and Power Company, Limited (Belgium v. Spain)* case, in which the Court recognised that:

> All the other States parties have a common interest in compliance with these obligations by the State in whose territory the alleged offender is present. That common interest implies that *the obligations in question are owed by any State party to all the other States parties* to the Convention. All the States parties *'have a legal interest'* in the protection of the rights involved. (Second Phase, Judgment, 1970: Para. 32–3; emphasis added)

In practice, this meant that any state could enjoy a legal standing before the Court when bringing a case against the state that perpetrated the breach of such a norm. In other words, the ICJ here distinguished obligations *erga omnes* from *jus cogens norms* in that the latter is necessarily part of the former – as all states can be said to have a legal interest in preventing a breach of *jus cogens* – yet the former might not qualify as *jus cogens.* Since the *Barcelona Traction* case, the ICJ has been evasive in setting out a clear and conclusive test as to what obligations qualify as *erga omnes,* but two key approaches have been identified: the material or the structural approach.

The material approach focuses on the relative importance of the rights involved. In the *Barcelona Traction* case, the ICJ emphasised this importance as such: 'In view of the importance of the rights involved, all States can be held to have a legal interest in their protection; they are obligations *erga omnes*' (Second Phase, Judgment, 1970, Para. 32–3). Although the court has been cautious in its use of *erga omnes* obligations since this judgment, it has on a few subsequent occasions reaffirmed it by invoking the ambiguous criteria of 'importance', in support of the material approach. This was for example the case in the 1996 *Application of the Convention on the Prevention and Punishment of the Crime of Genocide* case, where the Court asserted the prohibition of genocide as an obligation *erga omnes* due to its importance (Preliminary Objections, 1996; Para. 31). This seems in context that not only the State Parties, but also states not party to the Convention, ought to abide by this obligation due to the importance of such norms (Tams, 2005: 110). The obvious problem remains, however, that 'importance' remains an inherently vague and subjective notion. In applying the material approach to the CHM, the key issue would be to prove an equally high-level importance of preserving the CHM, which might be unrealistic considering the Court's scarce and careful case law to date.

The structural approach to obligations *erga omnes* goes beyond the linguistic emphasis of 'importance' and focuses rather on the logical nature of the obligation. This was most firmly elaborated on in the *Questions Concerning the Obligation to Prosecute or Extradite (Habré)* case, in which the Court established that the nature of the Convention against Torture vested in all states a common interest and thus obligations *erga omnes*:

> The common interest in compliance with the relevant obligations under the Convention against Torture implies the entitlement of each State party to the Convention to make a claim concerning the cessation of an alleged breach by another State party. *If a special*

interest were required for that purpose, in many cases no State would be in the position to make such a claim. It follows that any State party to the Convention may invoke the responsibility of another State party with a view to ascertaining the alleged failure to comply with its obligations *erga omnes partes,* such as those under Article 6, paragraph 2, and Article 7, paragraph 1, of the Convention, and to bring that failure to an end. (Judgement, 2012: Para 68–9; emphasis added)

This reasoning follows a strictly logical character, as it states that the common interest of all states in the compliance with the relevant obligations thus raises them to the level of *erga omnes partes.* The *Habré* case can be seen to have prompted this specific legal reasoning as an underpinning of the entirety of the functioning of the Convention against Torture. The ICJ followed a similar reasoning in the *Reservations to the Convention on the Prevention and Punishment of the Crime of Genocide* case in 1951:

In such a convention the contracting *States do not have any interests of their own;* they merely have, one and all, *a common interest,* namely, the accomplishment of those high purposes which *are the raison d'être of the Convention.* (Advisory Opinion, 1951: Para. 23; emphasis added)

This logical approach has as such been used in cases concerning *jus cogens* norms, but could potentially serve as an instrument for the legal interpretation of environmental obligations as well, as noted in the following observations by ICJ Judge Gaja:

On the logical plane, in the absence of such provision, the responsible State could avoid fulfilling any obligation of reparation when there is no injured State. *No State would in fact be able to invoke the responsibility of the wrongdoing State.* In the case of heavy pollution of the high seas or unlawful harm to the ozone layer, the responsible State would have an obligation of reparation that would not be owed to any other State and would therefore remain theoretical. This would also imply that the obligation not to pollute the high seas and not to cause harm to the ozone layer would also be theoretical, *for they could easily be breached without consequences.* (Gaja, 2010: 961; emphasis added)

Applying Judge Gaja's logic to the protection of the Area or any other component of the CHM, we can draw a similar theoretical conclusion, namely that if under the UNCLOS no state can prove to be injured for an unlawful appropriation of or unauthorised deep-sea mining in the Area, then in theory no state can bring a case against the responsible state. Even more concerning, by applying the same logic, the doctrine of CHM under the UNCLOS regime could be deemed redundant altogether, as no state could ultimately enjoy their rights to the Area, indeed the *raison d'être* of the common heritage of mankind.

Building a case under the UNCLOS

Regarding the avenues for dispute settlement under the UNCLOS, Article 287 sets out the different means regarding the interpretation or application of the Convention (UNCLOS 1982, Art. 287):

(a) the International Tribunal for the Law of the Sea established in accordance with Annex VI;
(b) the International Court of Justice;
(c) an arbitral tribunal constituted in accordance with Annex VII;
(d) a special arbitral tribunal constituted in accordance with Annex VIII ...

Article 288 thereafter establishes that these courts and tribunals shall have the jurisdiction over any such disputes regarding the interpretation and application of the Convention. It is interesting to note that neither Article 287 nor 288 specifies that the state initiating the procedure must defend its own individual interest. Thus, in reading the Convention as it currently stands, this could indeed allow for a state to defend its communal rights. What follows is that a disagreement about the interpretation of the UNCLOS can give rise to jurisdiction under Article 288, even if this disagreement would ironically concern the interpretation of the Convention's provisions in a state-centred or community-centred way. Hypothetically then, State A could initiate procedures against State B for breaking a provision that State A sees as establishing obligations towards all contracting parties, and thus pursuing it to defend its communal interest as it questions the interpretation and application of the Convention. Although no such case has yet been brought before the International Tribunal of the Law of the Sea (ITLOS), it remains a possible scenario.

As has been demonstrated, ITLOS and the ICJ have jurisdiction over the application and interpretation of the UNCLOS. In determining the avenues for protecting the CHM, it ought to be clear at the outset that by invoking *erga omnes* obligations, one does not alter the regime that traditionally gives rise to international responsibility. In the International Law Commission's Draft Articles on the Responsibility of States for Internationally Wrongful Acts (ASR), adopted in 2001, Article 1 states that 'Every internationally wrongful act of a State entails the international responsibility of that State', and in order to give rise to responsibility, the article requires that (i) the action or omission was attributable under international law, and (ii) constitutes a breach of an international obligation binding on it (ASR, 2001, Art. 1).

For the very purpose of clarifying the ambiguity of the scope of state responsibility and 'operationalising' the *erga omnes* concept, the ILC clarified that states could claim legal standing in case of an internationally wrongful act as an 'injured state' in Article 42:

Invocation of responsibility by an injured State

A State is entitled as an injured State to invoke the responsibility of another State if the obligation breached is owed to:
(a)   that State individually; or
(b)   a group of States including that State, or the international community as a whole, and the breach of the obligation:
  (i)   specially affects that State; or
  (ii)   is of such a character as radically to change the position of all the other States to which the obligation is owed with respect to the further performance of the obligation. (ASR 2001, Art. 42)

The difference between Article 42 (a) and (b)(i) seems rather ambiguous. The first part is clear in that the obligation was stipulated to the injured state individually, but the same could indeed be the case for (b) if the obligation is owed to all states. As to what 'specifically affected' in the sense of 42(b)(i) means, the ILC's commentary provides some guidance:

…a State may be specially affected by the breach of an obligation to which it is a party, even though it cannot be said that the obligation is owed to it individually (subparagraph (b)(i)). Thirdly, it may be the case that performance of the obligation by the responsible

State is a necessary condition of its performance by all the other States (subparagraph (b) (ii)); this is the so-called 'integral' or 'inter-dependent' obligation. (ASR 2001: 117)

Thus, if we are to read the UNCLOS as the treaty in question, Article 42 (a) could for example apply directly to State A in the case it would illegally conduct deep-sea mining operations in the Exclusive Economic Zone of State B. But the UNCLOS can also be seen as creating a so-called 'integral' obligation between all its State Parties, as under Article 139 it obliges all State Parties to 'ensure compliance and liability for damage'. Thus, should State A illegally conduct deep-sea mining operations in the Area, in such a scenario, State B would not only qualify as being 'specifically affected', but would also be under a duty to hold State A accountable for its violation of Article 137 of the UNCLOS (UNCLOS 1982, Art. 137 (1)). Should this argument be accepted, then all UNCLOS signatories could invoke the state at fault, either claiming to be injured or specifically affected; it would not even be necessary to argue standing *erga omnes*.

Should the state claiming to be injured or specifically affected not obtain legal standing under ASR Article 42, however, then the ASR Article 48 recognises some norms of 'higher importance' and that those obligations still can be invoked in the collective interest, meaning by states other than an injured state (Gaja, 2010: 961):

Article 48 *Invocation of responsibility by a State other than an injured State*

1. Any State other than an injured State is entitled to invoke the responsibility of another State in accordance with paragraph 2 if:
   (a) the obligation breached is owed to a group of States including that State, and is established for the protection of a collective interest of the group; or
   (b) the obligation breached is owed to the international community as a whole.

(ASR 2001, Supplement No. 10 (A/56/10),
UN Doc. A/56/10, Art. 48)

This recognises that if the obligations in question are 'collective obligations' established in a collective interest, other states than the injured state can intervene. As the ILC's commentary indicates, such collective obligations 'might concern, for example, the environment or security of a region' (ASR 2001, Commentary 1, Para 7). Thus, based on the wording in the preamble of the UNCLOS, the State Parties seem to declare that they sought to take 'into account the interests and needs of mankind as a whole' by designating the Area as CHM, and that 'the exploration of which shall be carried out for the benefit of mankind as a whole' (UNCLOS 1982, Preamble). It therefore seems reasonable to deduce that the obligation to protect the Area was indeed 'established in a collective interest'. One can also take the stance that any state, even those not party to the UNCLOS, can claim standing *erga omnes* in the collective interest being granted beyond the UNCLOS by way of customary international law (Wolfrum, 1983: 312–27).

Should a state ultimately be able to claim legal standing before a court 'on behalf of mankind' and claim reparations, the allocation of these remains a paradox. Should one invoke responsibility under ASR Article 48, Subparagraph 2 provides these possibilities:

> Any State entitled to invoke responsibility under paragraph 1 may claim from the responsible State:
> (a) cessation of the internationally wrongful act, and assurances and guarantees of non-repetition in accordance with article 30; and
> (b) performance of the obligation of reparation in accordance with the preceding articles, in the interest of the injured State or of the beneficiaries of the obligation breached.

If applied to a scenario where State A has unilaterally appropriated parts of the Area, all states would be the 'injured State' as the infringement on the Area implies that all states are equally injured by the loss of size of the Area. This is a general issue regarding *erga omnes* obligations, namely that no single state can in fact be designated as the beneficiary of the reparations. Thus, the possibility for seeking redress would become redundant, as the state invoking responsibility cannot claim it on its own behalf:

> In case of breaches of obligations under article 48, it may well be that there is no State which is individually injured by the breach, yet it is highly desirable that some State or States be in a position to claim reparation, in particular restitution. In accordance with paragraph 2 (b), such a claim *must be made in the interest of the injured State, if any, or of the beneficiaries of the obligation breached.* This aspect of article 48, paragraph 2, involves a measure of progressive development,

which is justified since it provides a means of protecting the community or collective interest at stake. (ASR 2001, Commentary re: Article 48, Para 12; emphasis added)

This remains a paradox, as Article 48 itself allows for invoking responsibility on behalf of the community but does not clarify to whom the respective reparation ought to be paid. To fill this legal lacuna, the establishment of a global fund that would be able to receive such compensation has been suggested (Scovazzi, 2005). This idea can be seen to have some support in the ILC's commentary to Article 37, as it lists 'a trust fund to manage compensation payments in the interests of the beneficiaries' as one of many possibilities of satisfactory compensation, but this has yet to be established (ASR 2001, Commentary re: Article 37, Para 5).

In 2011, the International Tribunal for the Law of the Sea's Seabed Disputes Chamber (SDC) issued an advisory opinion on *Responsibilities and Obligations of States Sponsoring Persons and Entities with Respect to Activities in the Area*, in which it touched upon many of the questions raised earlier in this chapter. It granted an interesting insight into the law pertaining to industrial activities in the Area, supporting the argument put forth above, namely that UNCLOS' environmental obligations are indeed norms *erga omnes partes* due to their intrinsic community interest:

> Each State Party may also be entitled to claim compensation in light of the *erga omnes* character of the obligations relating to preservation of the environment of the high seas and in the Area. (ITLOS Seabed Disputes Chamber, Case No. 17 (Advisory Opinion) 2011: Para 180)

In stating that the environmental obligations of the UNCLOS are obligations *erga omnes*, the SDC based itself on Article 48 of the ILC's Articles on State Responsibility. However, it remained as silent as Article 48 on the question regarding a potential beneficiary of remedies sought under this provision. Therefore, the question regarding the allocation of the reparations for harm to the Area is still unanswered to date. The question regarding reparations aside, this advisory opinion sets a significant precedent as, by recognising the existence of *erga omnes* obligations to preserve the Area, it solidifies the view that legal scholars have argued for based on the structural approach (Johnstone, 2015: 6). It can be seen as a step towards opening avenues for legal action based on the communal interest in preserving the Area as common heritage of mankind.

Conclusion

Having brought together the scholarly literature on the CHM designation with the legal commentaries on the *erga omnes* doctrine,
this chapter has attempted to shed light on the possible avenues
of redress should the Area be exploited in an unlawful manner. In
the scarce case law in which the ICJ has invoked obligations *erga
omnes*, these have been mentioned in conjunction with an emphasis
on the 'importance' of the norm at stake; hence the Court seems to
largely base itself on the material account. In light of this careful
jurisprudence, it seems unlikely that the protection of the Area will
be raised to the same level of importance as preventing genocide
and torture, at least in the nearest future.

Although the protection of the Area might not qualify as an
obligation *erga omnes* before the ICJ, this chapter has still found
that the ILC's Articles 42 and 48 on State Responsibility could be
granting legal standing based on community interest. Both Article
42 and Article 48 could indeed be found to grant legal standing to
a State Party to the UNCLOS in case of harm inflicted on the Area,
and should this be the case, then *erga omnes* obligations would not
even be necessary to invoke responsibility. The key difference is however that should a state that is not party to the UNCLOS exploit the
Area unlawfully, then it would not be possible to prosecute them as
being in breach of the Convention, as opposed to the case where the
obligations would be entrenched in international law as obligations
erga omnes.

This interpretation follows Judge Gaja's logical account, in which
light we can argue that without recognising an *erga omnes* standing
for the CHM, the *raison d'être* of the designation is obliterated
(Gaja, 2010: 961). This follow the structural approach; if the Area
belongs to everyone, even states not party to the UNCLOS, then all
states can be said to be beneficiaries of a continued, integral, compliance with the obligations protecting it. If the legal framework
would grant no one state legal standing should a violation occur,
then the obligation becomes merely theoretical and in fact nobody
will have an incentive to comply, which in turn would render the
designation of the CHM redundant as no state would in fact be able
to invoke the responsibility of the wrongdoing state.

The final advisory opinion of the ITLOS' Seabed Disputes Chamber (SDC) is supportive of this approach yet leads to another dead-end regarding reparations. To this day, it remains unclear who could be the recipient of the reparations for said hypothetical harm inflicted on the Area. It seems that both the SDC and the commentaries to the ASR keep deliberately silent to this end, and only time and jurisprudence will be able to provide any answers.

The final finding is that the UNCLOS itself does not explicitly require states to defend *individual* rights in invoking jurisdiction under Articles 278 and 288. This leaves a door open for a hypothetical case to be brought by a State Party to the UNCLOS, aiming to act in the collective interest as it initiates a case regarding the interpretation and application of the Convention. Time will show if such a case will ever be brought. What has been demonstrated throughout this chapter is then that the legal framework for effectively protecting the Area by allowing 'States other than an Injured State' to initiate proceedings may indeed be available to us if a relevant case would arise and the will for such an interpretation of the law prevails in the international courts.

Although the doctrines of *erga omnes*, *jus cogens*, CHM, and other community-centred concepts have irreversibly entered the legal toolbox and we have seen an increasingly progressive approach to indeed invoke these doctrines, it is fair to note that they have been met with considerable scepticism (Tams and Asteriti, 2013: 164). A classic positivist argument invoked to rebuke the likelihood and desirability to expand the *erga omnes* doctrine is based on the doctrine *pacta tertiis nec nocent nec prosunt*, which translates as 'a treaty cannot create obligations binding upon third states without their consent' (Pauwelyn, 2003: 102–3). This doctrine, being deeply rooted in public international law, has been reflected in the cautious approach by the ICJ in invoking the doctrine since the *Barcelona Traction* case. In his dissenting opinion in the *Nuclear Tests (Australia v France)* cases, ICJ Judge de Castro noted that the concept of obligations *erga omnes* as expressed in the *Barcelona Traction* case should be taken '*cum grano salis*':

> it seems to me that the *obiter* reasoning expressed therein should not be regarded as amounting to recognition of the *action popularis* in international law; it should be interpreted more in conformity

with the general practice accepted as law. I am unable to believe that by virtue of this dictum, the Court would regard as admissible, for example, a claim by State A against State B that State B was not applying 'principles and rules concerning the basic rights of the human person' (ICJ Reports 1970, p. 32, para. 34) with regards to the subjects of State B or even C. (Dissenting Opinion of Judge Castro (1974) ICJ Rep 372 at 387; 34)

This scepticism is indeed one that is reflected in the scarcity of case law and clear, progressive development with regard to the *erga omnes* doctrine. However, in the fifty years that have passed since the *Barcelona Traction* case, a well of scholarly literature and debates regarding the concept have been kept alive in academia. Thus, while still respecting the reluctance by judges and scholars that raise their concerns for the coherence of the international legal order, this chapter is but one in many works that have explored its potential applicability with regard to contemporary issues such as deep-sea mining.

The legal regime for the common heritage of mankind at sea is currently fragmented and ambiguous, and the level of enforcement seems unable to manage the emerging challenges and activities on the deep seabed following the surging interest in deep-sea mining. Regarding the environmental impact of such activities, many questions remain unanswered as large-scale deployments of mining operations are just starting to materialise. This chapter has explored how the concept of obligations *erga omnes* could be adopted to enhance compliance when states now rally to control and dominate the maritime resources located on the deep seabed. Since it was introduced in the second half of the last century, it has been applied in a very careful manner and thus merely waded the waters of its potential range of applicable spheres. This chapter has preliminarily found that if the CHM designation ought to live up to its promise and ensure that current and future generations benefit equally from the Area, then the *erga omnes* doctrine could be used to empower states to defend their rights to this end. Ultimately, such a legal instrument could help to balance the power politics that are currently playing out with regard to the deep seabed, and as such influence international relations for many years to come. If obligations *erga omnes* are ultimately recognised and interpreted as such, it could indeed prevail as a powerful tool in the legal toolbox, used to ensure that the common heritage of mankind remains neither an apology nor a utopia in a Koskenniemian sense, but rather something closer to reality (Koskenniemi, 2006).

References

Anand, R. P. (1997). 'Common heritage of mankind: mutilation of an ideal', *The Indian Journal of International Law* 37: 1–18.

Borgese, E. M (1995). *Ocean Governance and the United Nations*, 2nd edn, Halifax, Nova Scotia: Centre for Foreign Policy Studies, Dalhousie University.

Cruickshank, M. J. (2016). 'Ocean raw materials', *Kirk-Othmer Encylopedia of Chemical Technology*, https://onlinelibrary.wiley.com/doi/book/10.1002/0471238961 (accessed 12 April 2022).

Egede, E. (2014). 'Common heritage of mankind', in A. Carty, ed., *Oxford Bibliographies Online: International Law*, Oxford: Oxford University Press.

Gaja, G. (2010). 'The concept of an injured state', in J. Crawford, A. Pellet and S. Olleson, eds, *The Law of International Responsibility*, Oxford: Oxford University Press, 943–7.

Gaja, G. (2014). 'Protecting community interests concerning natural resources', in *The Protection of General Interests in the International Community*, vol. 364, Leiden: Brill, ch. 14, 171–80.

Grotius. H. (1916 [1608]). *Mare Liberum or The Freedom of the Seas or the Right Which Belongs to the Dutch to Take Part in the East Indian Trades*, trans. R. Van Deman Magoffin, New York: Oxford University Press.

Hardin, G. (1968). 'The tragedy of the commons', *Science* 162(3859): 1243–8.

ISA Exploration Contracts (2022). 'Exploration contracts', https://isa.org.jm/exploration-contracts (accessed 12 April 2022).

ISA Press Release (2021). 'Nauru requests the President of ISA Council to complete the adoption of rules, regulations and procedures necessary to facilitate the approval of plans of work for exploitation in the Area', www.isa.org.jm/index.php/news/nauru-requests-president-isa-council-complete-adoption-rules-regulations-and-procedures (accessed 12 April 2022).

Johnstone. R. L. (2015). 'Invoking responsibility for environmental injury in the Arctic Ocean', *The Yearbook of Polar Law* 6(1): 1–35.

Kasa, S., A. T. Gullberg and G. Heggelund (2008). 'The Group of 77 in the international climate negotiations: recent developments and directions', *International Environmental Agreements: Politics, Law and Economics* 8(2): 113–27.

Koskenniemi, M. (2006). *From Apology to Utopia: The Structure of International Legal Argument*, Cambridge: Cambridge University Press.

Pauwelyn, J. (2003). *Conflict of Norms in Public International Law: How WTO Law Relates to Other Rules of International Law*, Cambridge: Cambridge University Press.

Pinto, M. C. W. (1996). "Common heritage of mankind"': from metaphor to myth, and the consequences of constructive ambiguity', in J. Makarczyk, ed., *Theory of International Law at the Threshold of the 21st Century: Essays in Honour of Krzystof Skubiszewski*, Cambridge: Cambridge University Press, 249–68.

Pinto, M. C. W. (1983). 'The new law of the sea: the lawyer as international legislator', in P. Van Dijk, ed., *Restructuring the International Economic Order: The Role of Law and Lawyers*, Boston, MA: Kluwer.

Platzoder, R., ed. (1987). *12 United Nations Conference on the Law of the Sea: Documents 302*, New York: United Nations Publication.

Posner, E. (2008). '*Erga omnes* norms, institutionalization, and constitutionalism in international law', *Public Law and Legal Theory Working Paper No. 224*, Chicago: University of Chicago Law School.

Scholtz, W. (2008). 'Common heritage: saving the environment for humankind or exploiting resources in the name of eco-imperialism?' *The Comparative and International Law Journal of Southern Africa* 41(2): 273–93.

Scovazzi, T. (2005). 'Some remarks on international responsibility in the field of environmental protection', in M. Ragazzi, ed., *International Responsibility Today: Essays in Memory of Oscar Schachter*, Leiden: Brill, 209–22.

Tams, C. J, and A. Asteriti (2013). 'Erga omnes, jus cogens, and their impact on the law of responsibility', in M. Evan and P. Koutrakos, eds, *The International Responsibility of the European Union*, Oxford: Hart Publishers, 163–88.

Tams, C. J (2005). *Enforcing Obligations* Erga Omnes *in International Law*, Cambridge: Cambridge University Press.

Tuerk, H. (2012). *Reflections on the Contemporary Law of the Sea*, Leiden: Martinus Nijhoff.

UNCLOS (1982). 'United Nations Convention of the Law of the Sea', www.un.org/depts/los/convention_agreements/texts/unclos/unclos_e.pdf (accessed 12 April 2022).

Wolfrum, R. (1983). 'The principle of the common heritage of mankind', *Heidelberg Journal of International Law* 43: 312–37.

Wolfrum, R. (2011). *Enforcing Community Interests Through International Dispute Settlement: Reality or Utopia?* Oxford: Oxford University Press.
Young, O. R. (1989). *International Cooperation: Building Regimes for Natural Resources and the Environment*, Ithaca, NY: Cornell University Press.

Case law

Application of the Convention on the Prevention and Punishment of the Crime of Genocide, Preliminary Objections, Judgment, ICJ Reports (1996).
Barcelona Traction Light and Power Company Limited (Belgium v Spain) [1970] Reports (ICJ).
Nuclear Tests (Australia v France) (Dissenting Opinion of Judge Castro) (1974) ICJ Rep 372.
Questions Concerning the Obligation to Prosecute or Extradite (Belgium v. Senegal), Judgment (20 July 2012).
Reservations to the Convention on the Prevention and Punishment of the Crime of Genocide, Advisory Opinion, ICJ Reports 1951.
Responsibilities and Obligations of States Sponsoring Persons and Entities with Respect to Activities in the Area, International Tribunal for the Law of the Sea: Seabed Disputes Chamber, Case No. 17 (Advisory Opinion), 1 February 2011.

Legal instruments

International Law Commission, 'Draft Articles on Responsibility of States for Internationally Wrongful Acts 2001, with commentaries', *Yearbook of the International Law Commission*, 2001, vol. II, Part Two, 117.
International Law Commission, *Draft Articles on Responsibility of States for Internationally Wrongful Acts*, November 2001, Supplement No. 10 (A/56/10), UN Doc. A/56/10.
The Vienna Convention on the Law of Treaties (the Vienna Convention) (Vienna, 23 May 1969, 1155 UNTS 331). Reprinted in 1969: *International Legal Materials* 8: 679.
UN Doc. A/Conf. 62, GP 9, 5 August 1980.
UN Doc. A/Conf. 62/L. 58, XIV UNCLOS, 129. Report of the President on the Work of the Informal Plenary Meeting of the Conference on General Provisions, 1982.
UN General Assembly Res. 2749 (XXV), para 1, UN Doc. A/RES/25/2749, 12 December 1970.
UN General Assembly Res. 3201-S VI of May 1974.
United Nations Convention on the Law of the Sea, 1833 U.N.T.S. 397, 10 December 1982.

9

Fishing for territory: historical International Relations and the environment

Kerry Goettlich

The herring is one of those products whose use decides the destiny of empires. The coffee bean, the tea leaf, the spices of the torrid zones, the worm which spins silk, had less influence on the wealth of nations than the northern ocean.

– Bernard Germain de Lacépède (1756–1825),
French naturalist and politician[1]

International Relations (IR), as some have argued, is sea-blind. In many of the key IR theory texts, seas and oceans are rarely mentioned, and when they are, they appear mostly as empty space that has to be traversed (Waltz, 1979; Mearsheimer, 2001). In part, this conspicuous absence of most of the world's surface should be seen in the context of a larger, somewhat enduring absence of the sea in the social sciences, which will likely continue to be produced mainly by people writing from places on dry land. The early twentieth-century geographer Ellen Semple was not alone in noticing that 'Our school textbooks in geography present a deplorable hiatus, because they fail to make a definitive study of the oceans', and that 'no history is entitled to the name of universal unless it includes a record of human movements and activities on the ocean' (quoted in Steinberg, 2001: 9).

This chapter, however, approaches that absence from a different angle. While the unique role of maritime environments, it could be argued, is absent from IR at a general level, it is not clear how much more present land itself is than sea, or for that matter, air, outer space, or any other particular environment. The way in which IR is largely based on abstract concepts intended to see beyond, or even obscure, the significance of particular environments is not limited to its abstraction from the ocean (Barkawi, 2009). Of course, struggles over territory are prominent in disciplinary narratives, but discussion of what territory itself is and how it is produced out of land remains limited (although see Elden, 2010). In historical IR and in historical sociology, territoriality is often confused with the autonomy of a sovereign state, and the changing ways in which boundaries have or have not been surveyed and marked within landscapes have not usually been seen as significant.[2]

In reconceptualising the role of the ocean in IR, then, it matters whether we see the absence of the ocean as a problem in itself, or whether it is a symptom of some larger feature of the discipline. This chapter, in contributing to attempts to bring the ocean into IR as a distinct environment of international politics, uses this attempt as an opportunity to advance a larger effort in IR and the broader social sciences to take the environment seriously. The need for new ways of analysing the connections between humans and the environment has in many ways been reflected in a wealth of innovative IR scholarship since the 1990s.[3] Research in this area has ranged from perspectives that add in environmental degradation as a new concern in an already established security framework to those that challenge the distinction between humans and nature.

At the same time, increasing awareness of the agency of humans on a geological scale has also come with limitations. The prevailing scholarly emphasis on climate change and comparable environmental catastrophes, along with their political implications, is understandable within the context of the present historical moment. With this emphasis, however, often comes a focus on emerging environmental threats and relatively recent or potential efforts to address them (O'Neill, 2009; Fishel et al. 2016; Chasek et al., 2018). There have been efforts within the environmental security literature to critique this tendency, leading to a richer understanding of security and the environment. Simon Dalby, for example, argues that leaving environmental history out can lead to overestimating both the regularity of past weather patterns and the novelty of humans' environmental concerns (Dalby, 2002: 69–73). Neo-Malthusian accounts stressing recent economic scarcity and overpopulation, he argues, draw attention away from the longer-term political conditions of environmental degradation. A closer engagement with history, then, would improve understanding of the nature and causes of environmental degradation and its political consequences.

Historians themselves, however, may have different ideas about the uses of environmental history. Dipesh Chakrabarty, for one, approaches climate science as not just an area of knowledge that can be enriched by a historian's perspective, but also the other way around, because awareness of humans as geological agents makes any distinction between human and natural history untenable (Chakrabarty, 2009b). As Chakrabarty recounts:

> As the [climate] crisis gathered momentum in the last few years, I realized that all my readings in theories of globalization, Marxist analysis of capital, subaltern studies, and postcolonial criticism over the last twenty-five years, while enormously useful in studying globalization, had not really prepared me for making sense of this planetary conjuncture within which humanity finds itself today. (Chakrabarty, 2009a: 25)

Depending on our object of analysis, existing categories may be sufficient. But once climate change comes into view, practically all history changes profoundly, and historians of modernity and globalisation, willing or not, become implicated in questions of natural history. Enlightenment ideas of human agency, for example, now appear differently when juxtaposed with the geological agency of humans, which may be 'the price we pay for the pursuit of freedom'. 'A critique that is only a critique of capital', to take another category of human history, 'is not sufficient for addressing questions relating to human history once the crisis of climate change has been acknowledged' (Chakrabarty, 2009b: 212).

Chakrabarty's reevaluation of historical knowledge has important consequences for historical IR. Instrumentalising particular historical knowledge towards a better understanding of future catastrophes, while important on its own, would not necessarily affect our understanding of history itself. But it is not just particular historical knowledge about nature but the idea of human history itself which is challenged by the geological agency of humans. What, then, would historical IR look like if it not only took into account nature as natural resources or as a social construction, but also took seriously the problem of separating out natural history itself from the history of the international? How would our histories of the core categories of disciplinary IR, such as power, sovereignty, or territory change if it was no longer assumed that 'natural' entities had no role to play?

This chapter explores these questions by focusing on the role of fish in international relations, in two different ways. The first section takes a broad global historical perspective, making the case that fish have played a strong role in influencing the direction of maritime empires' development. In contrast to many accounts in which maritime empire or sea power is largely dependent on land-based phenomena, for example through trade with terrestrial societies, control of the sea has in many cases historically been sought after in pursuit of the sea's own contents. The second section makes a more specific argument about the place of fish in the global history of territoriality, examining scientific debates about overfishing, from the late nineteenth century onwards. Overfishing was initially shown by the philosopher of science Thomas Huxley to be impossible, but this conclusion was overturned by later scientists, leading states to reverse a long-standing international legal principle and claim exclusive fishing areas. The current territorialisation of the ocean, then, is to a significant degree an outcome of human interactions with fish.

By devoting attention in new ways to the international politics of human–environment relations, the chapter also engages with the themes for studying the ocean put forward in this volume's introduction (de Carvalho and Leira, 2022), particularly those of the sea as a wilderness to be tamed, and the sea as a resource. While the environment is often seen as a series of resources to be extracted to further human productivity, the chapter stresses the ways in which ocean resources, particularly fish, have historically been politicised. Thus the trope of the sea as a resource blurs into that of the sea as a wilderness which, with more or less difficulty, can be dominated. While useful as a source of nutrition, fisheries are also a basis on which maritime empires can be built in numerous other ways, and it will be seen that even attempts by people, scientific or otherwise, to understand the movements and cycles of fish well enough to catch them are deeply political.

Piscine empire

In 1620 a group of English religious separatists later known as 'Pilgrims' landed at modern-day Plymouth, Massachusetts, and began to starve, freeze, and die of scurvy in the New England winter (Adolf, 1964: 249). After some months, an Abenaki sagamore called Samoset came into their camp and greeted them in English. According to some accounts he asked the English if they had any beer (Davey, 2018: 9). Samoset introduced them to Tisquantum, also an English speaker, who despite having previously been captured and sold into slavery by an Englishman, would serve as a key intermediary with and enabler of the nascent English settlement.[4] Samoset knew English, and perhaps had a taste for English beer, because English fishers had been fishing in the area alongside Native Americans and various other Europeans since the fifteenth century. Indeed, there were so many fish in the area that one explorer noted that in a few hours of fishing, he and his companions 'had pestered our ship so with Cod fish, that we threw numbers of them over-boord againe' (Cronon, 2003: 22). While early New England, from its beginnings at 'Cape Cod', was heavily dependent on fishing for its survival, the English colony further north in Newfoundland consisted of little else, its formal creation being an attempt to bring royal protection to a fishing colony already in existence.

Maritime empires often have some particular relationship with fish and other sea life. For example, China's claim over the South China Sea hangs in no small part on its assertion that Chinese fishers have historically derived rights over fisheries there (Gao and Jia, 2013).[5] The 'truce' referred to in the name 'Trucial States', a confederation of Persian Gulf peoples organised within the British Empire from 1820 to 1971, specifically protected the pearling season, which was the basis of the Gulf economy (Onley, 2009). In the 1890s, a Russian Far Eastern fishing industry was created from scratch by the Russian government by offering land, equipment and loans to Baltic and Finnish fishers, mainly in order to wrest control over its eastern seas away from Chinese, Korean, and Japanese fishers (Sokolsky, 2015). Maritime empires that are sustained or motivated by fishing, then, might be called 'piscine empires'.

While naval war comes and goes, with many historical polities not maintaining a standing navy, demand for fish is almost always more consistent. As a result, fishing in many contexts has been a training ground for naval recruitment, and provided a reliable pool of potential seamen. Britain and France both considered the Newfoundland fisheries in the eighteenth century to be a vital strategic interest for this reason, and they were a major issue in peace negotiations at Utrecht (1713), Paris (1763), and Versailles (1783) (Hiller, 1996). A quarter of the men available for service in the French navy came from the Newfoundland fisheries, which was believed to create superior mariners, '*pur et dur*', through its particular hardships. Today China maintains a naval militia made up of fishing boats, which have been accused of ramming Vietnamese fishing boats in the South China Sea (Sevastopulo and Hille, 2019). Many Vietnamese fishers understand themselves, in consequence, as defending Vietnamese sovereignty even as they earn their livelihood.

One way of illustrating the significance of fish in early modern European maritime empires is through the international legal debates that set parameters for them beginning in the seventeenth century. Under papal sponsorship Portugal and Spain had declared extensive spheres of influence, including vast ocean spaces. The Dutch lawyer Hugo Grotius was among many outside Portugal and Spain seeking to delegitimate this Iberian hegemony, and argued, in brief, that the seas were essentially open to all. A key part of his argument was that the oceans could not be appropriated because their use through fishing could never be exhausted (Anand, 1983: 81). In this formulation, the notions of the sea as an unlimited resource and as an uncontrollable wilderness work together; if one could never exhaust its resources, one could not dominate it. Grotius extensively references ancient Roman fishing law in order to show the long history of the freedom of the sea. For example, he distinguishes between the sea proper and the channels sometimes dug inland from the sea by wealthy Romans to create private saltwater fish ponds, called *diverticula* (Ziskind, 1973: 543; Grotius, 2004: 28). Similarly, he cites the Roman lawyer Ulpian as deeming it illegal to prevent someone from fishing in front of one's house (Ziskind, 1973: 542; Grotius, 2004: 29).

At the time *Mare Liberum* was published in 1609, pervasive Dutch fishing was a recurring English grievance against the Dutch. According to one economic historian, fishing was 'as much the base of the Dutch economy as agriculture was of the English', and the

Dutch herring fleet with its several thousands of ships followed herring shoals annually from the Shetland islands down to the Thames estuary (Wilson, 1967: 520). Thus William Welwood, a Scottish lawyer, argued that Grotius' intention in writing *Mare Liberum* was really to curb England's right to control its own fisheries, and that Grotius' attacks on the Portuguese East India trade monopoly was only a convenient excuse (Fenn, 1926: 175). Navigation was 'a thing farre off from all controversie, at least upon the Ocean', whereas the use of fisheries was a more realistic motive (Welwood, 2004: 65). Welwood noted that it was incorrect to assume that the ocean's resources were limitless, as in fact the Dutch had been systematically overfishing:

> [I]f the uses of the seas may be in any respect forbidden and stayed it should be chiefly for the fishing ... For whereas aforetime the white fishes daily abounded even into all the shores on the eastern coast of Scotland, now forsooth by the near and daily approaching of the buss-fishers the shoals of fishes are broken and so far scattered away from our shores and coasts that no fish now can be found worthy of any pains and travails, to the impoverishing of all the sort of our home fishers and to the great damage of all the nation. (Welwood, 2004: 74)

As such, Welwood thought that the logic of *Mare Liberum* itself suggested that for the sake of conservation, England should be able to regulate the fisheries, in order to enforce minimum mesh requirements for nets and allow stocks to regenerate (Edmond, 1995: 208). Moreover, he accepted the link made by Grotius between the exhaustibility of the sea's resources and the capability of polities to control maritime spaces. Ultimately John Selden's *Mare Clausum* would become the more famous English defence against Grotius, written when an Anglo-Dutch fishing conference was triggered by English ships capturing some Dutch ships off of Greenland, laden with twenty-two walruses (Fletcher, 1933: 8). But both Selden and Welwood took fishing rights seriously as some of the main material with which Grotius's historical claims could be dismantled, referring to Grotius's examples of the Roman *diverticula* and the prohibition against preventing someone from fishing outside one's house (Ziskind, 1973: 547; Welwood, 2004: 68).

The United States in its exclusive economic zone (EEZ) remains a piscine empire today, keeping strict controls on foreign fishing, with its EEZ made particularly extensive by its many island dependencies (Kraska, 2011: ch. 7). But in general piscine empire today is best understood as dispersed across the world's oceans, wherever the major fishing fleets go. One common phenomenon today that can be understood as piscine empire is the creation of access agreements between major distant-water fishing (DWF) states and states in the Global South (Gagern and van den Bergh, 2012). These agreements first appeared in the 1970s, shortly after the creation of the EEZ through UNCLOS III. While the newly independent postcolonial states had managed to take economic control of these maritime zones, large trawlers usually belonged to fleets in the Global North, and these had a much greater fishing capacity. Global fish catches peaked in the late 1980s and have been declining since then, despite increasing fishing activity, showing a general problem of overfishing (Pauly et al., 2002: 691). In this context, primarily the EU, Japan, and Russia began negotiating agreements to fish in the EEZs of Global South states, aiming to meet rising demand, with China joining later on.

While access agreements have been concluded virtually all across the latitudes nearest to the equator, West Africa is a major focus of these agreements. In this region, from Morocco to Namibia, the EU and China represent the main DWF fleets. Agreements made with the EU usually consist of a sum of money paid to the host government, along with other stipulations such as requirements to land a certain amount of fish in the host country and to provide support for technological and industrial development (Belhabib et al., 2015). In exchange, EU vessels receive the right to fish in various designated areas within the host country's EEZ, according to the type of fish caught and equipment used. Chinese agreements tend to be more secretive, but seem to combine payments with more direct expenditures on infrastructure, military equipment, and debt relief.

While the creation of the EEZ in the 1970s seemed to promise a kind of decolonisation of ocean spaces after a long period of Anglo-American maritime hegemony, these Chinese and European fishing zones have generally prevented West Africans from experiencing benefits from it (Iheduru, 1995). According to one estimation, fishing access in West Africa from 2000 to 2010 cost the EU 26 per cent of reported landings, and China spent 40 per cent of the value of its reported landings on access agreements (Belhabib et al., 2015: 10).

But because fishing is very difficult to monitor, including unreported landings brings the figures down to 8 per cent for the EU and 4 per cent for China. In exchange for these payments, European and Chinese trawlers have overfished or severely depleted many West African fisheries, making it difficult for more local fishers to catch fish. One recent study indicated a decline in biomass by a factor of 13 in Northwest African fisheries since the 1960s, with large trawlers picking up vast amounts of unwanted organic material (Alder and Sumaila, 2004: 169). In many cases, local fishers have become poorer despite the value of fisheries increasing, with DWF fleets rarely landing in West African ports at all, and much food is imported (Kaczynski and Fluharty, 2002). Ghana, which was at its independence in the 1950s a regional fishing power sending substantial fleets to Senegal and the Congo, has declined largely as a result of foreign overfishing (Atta-Mills and Sumaila, 2004). While the EU employs rhetoric of sustainable development and has set up mechanisms for consultation with West African interest groups and civil society representatives, little has been done to address the basic imbalance between large subsidised trawlers and small artisanal fleets (Antonova, 2016).

The political implications of distant-water fishing, moreover, have not been limited to coastal states where food sources have been threatened. In particular, the piscine imperialism of DWF fleets has had a crucial role in the recent history of piracy in the Indian Ocean. In waters near Somalia, distant-water fishing has been particularly intense since Somalia's civil war began to leave its governing authority in question (Kamola, 2018). The UK and Norway, for example, have challenged Somalia's claim to an EEZ because it was not filed properly with the UN, in theory leaving the waters near Somalia open to unlimited fishing (Hansen, 2011: 29). Many Somali pirates claim to be simply fighting foreign intruders, and this claim resonates with many Somalis beyond pirates (Dagne, 2009: 105–6). Whether or not foreign fishing is an economic 'cause' of piracy, and the merits of seeing the pirates as 'Robin Hood' figures simply defending their shores, have been much debated. But 'defensive piracy' did exist to some extent among the former Somali coast guard following the state collapse in 1991, and proponents and critics of these narratives largely agree that foreign fishing has played an important role by giving pirates a social or political context (Samatar et al., 2010: 1385). On one level, the narrative of pirates as a Somali coast guard provides coherence and meaning to the pirates themselves (Bueger, 2013). But beyond this, as some Somalis have pointed out, the inequities of foreign fishing create a strong reason to view with irony or indifference the international attention given to the issue of piracy, and many 'see the discourse on piracy as a clear manifestation of the double standards used in the international system' (Samatar et al., 2010: 1389).

Fisheries conservation science and the bounding of the oceans

Fish, then, can potentially play a crucial part in sustaining and motivating maritime empire. More generally, moreover, attention to relations between humans and non-humans is important for IR not just in terms of the threats that environmental degradation pose to the human species at large, and that these relations with non-humans can end up defining important aspects of international politics. But is a better understanding of fish really necessary in IR? In order to argue that it is, this section demonstrates the crucial role of human interactions with fish in the way territory extends into ocean space. In particular, human–fish relations were at first central to efforts of the late nineteenth century to avoid state regulation of maritime space beyond the traditional three-mile limit, and then later to the creation of the (EEZ), one of the basic legal concepts which govern the ocean.

Fish and fisheries do not lend themselves easily to linear boundaries (Meltzer, 1994). To the extent that fish form part of an apparent marine wilderness which is unpredictable and difficult to pin down, they are often not as easy to territorialise as land resources. Fish have been territorialised informally through customary rules in a number of different contexts, including Tokugawa Japan and to this day in the Solomon Islands and in Maine and Canada, but on a limited scale, and not with fixed, linear boundaries (Durrenberger and Palsson, 1987; Aswani, 2005). But due in part to historical changes in fisheries science, I argue, it became possible for states to begin claiming well-defined ocean territories.

Fisheries science in its current form began in the nineteenth century as a response to fishers' claims that they had observed overfishing (Hubbard, 2011: 11). Before this time it was not always clear whether the fluctuations that had been observed in fish populations were directly caused by fishing. In medieval Europe, reports of overfishing of freshwater fish were commonplace, with Philip IV of France declaring in 1289 that 'every river and water-side of our realm, large and small, yields nothing due to the evil of the fishers and the devices of [their] contriving, and because the fish are prevented by them from growing to their proper condition' (Hoffmann, 1996: 648). Yet Grotius, as noted above, supported the concept of *Mare Liberum* with the notion that overfishing was not possible. For Emer de Vattel, moreover, 'It is manifest that the use of the open sea, which consists in navigation and fishing, is inno-cent and inexhaustible; that is to say—he who navigates or fishes in the open sea, does no injury to anyone' (de Vattel, 2008: 250).

This ambiguity around the question of overfishing persisted through the period of the industrial revolution. Steam trawlers were popularised among British fishers, as they caught many more fish by actively dragging a large net through the water rather than leaving it in place. Traditional line fishers then complained that trawlers were depleting stocks, and the UK government appointed a commission to investigate in 1863–66. The commission was chaired by Thomas Huxley, who was known as 'Darwin's bulldog' for his major public role in British philosophy of science debates, and was involved in many royal commissions on topics from edu-cation reform to vivisection (White, 2003: 67). The commission was charged with investigating changes in the supply of fish and the likely effects of regulating the fishing industry (in Levi, 1866: 184–214). It visited and gathered evidence from eighty-six fishing locations across Britain, Ireland, and the Isle of Man, but the evi-dence that mostly guided its conclusions were public records. It found no 'direct' evidence on the overall supply of fish, but the fish carried by the main railroads had increased threefold in nine years, while the overall price of fish had been increasing, but not faster than that of butchers' meat.

For the commission, this suggested that the supply of fish was increasing, and had not been negatively affected by overfishing. The commission consulted fishermen all across the UK but on the whole did not consider the information gained in this way to be useful. Because different groups of fishermen depended for their livelihoods on different fishing methods, it surmised,

> it will not be a matter of astonishment that the evidence, so far as it records merely personal convictions, and assertions that can neither be proved nor disproved, is of the most conflicting character … fishermen, as a class, are exceedingly unobservant of anything about fish which is not absolutely forced upon them by their daily avocations…
> (in Levi, 1866: 189)

The commissioners were confronted with conflicting data. When fishers did suggest that there was overfishing, the commission registered this with scepticism, noting that they were

> not only prone to adopt every belief, however illfounded, which seems to tell in their own favour, but they are disposed to depreciate the present in comparison with the past. Nor, in certain localities, do they lack the additional temptation to make the worst of the present, offered by the hope that strong statements may lead the State to interfere in their favour, with dangerous competitors (in Levi, 1866: 189)

There was no such scepticism, however, about statistics showing that railroads were transporting more and more fish inland. As a result, the commission concluded that the supply of fish was increasing, that fishing methods such as beam trawling were not wasteful, and that any general regulations on fishing would be harmful to the UK food supply. Later, Huxley restated these convictions in an 1883 speech inaugurating the International Fisheries Exhibition in London: 'probably all the great sea fisheries, are inexhaustible; that is to say, that nothing we do seriously affects the number of the fish. And any attempt to regulate these fisheries seems consequently, from the nature of the case, to be useless' (quoted in Smed and Ramster, 2002: 366). This speech, and the inquiry from which it emerged, is often cited as the main impediment to regulations on trawling or the taking of immature fish in the UK in the late nineteenth century.

It was not a coincidence that the first fishing inquiry of its kind all but ignored the first-hand accounts of overfishing that had motivated its formation. Huxley's role in the fisheries commission, along with many other commissions, formed part of his effort to create a new public figure he called the 'man of science', who would be all but immune to politics and society (White, 2003: ch. 3). Against the 'superstitions' and special interests of the aristocratic literary elite, dominated by the Church of England, and at the same time against 'popular' science writers aiming at amusement and profit, Huxley wanted to create an amateur public audience. Although this would to some extent democratise knowledge production, by enabling laboratory practitioners to become public intellectuals alongside literary figures, it would also serve a national interest and depoliticise the monopolisation of knowledge by experts, whose expertise would be confirmed by an amateur audience. Huxley's well-known public defence of the country squire Charles Darwin, for which he was called 'Darwin's bulldog', was based in a valuation of genius, independent from patronage and political ambitions, in seclusion from society, whether on a transoceanic voyage, in a country home, or in the laboratory (White, 2003: 60).

The fisheries commission was motivated by this sense of the role of science in society. While the national interest as a whole was at stake, in particular in terms of the food supply for a rapidly growing population, the 'man of science' had to remain strictly isolated from particular interests within society. Conflicts of interest between groups of fishers using different fishing methods meant that all their testimony had to be seen as squabbles between particular interests trying to discredit each other. The data collected on fish prices and the amount of fish being transported, on the other hand, were verifiable facts, and formed apparently solid grounds for assessing the effects of trawling. The commission's conclusions, then, came out of a desire to separate the natural sciences from politics. Yet this was, of course, highly political, in that it assumed, as the ultimate goal, a unified national interest that could be isolated from particular group interests, and based its reasoning on the measurement of the numbers of available fish on these political principles.

As long as there was no rationale for state intervention into fishing practices, there was little impetus for the territorialisation of the sea. Yet Huxley's attempts to shed the light of rational science on the mysteries of fish fluctuations did not settle the debate. In the early twentieth century the debate on overfishing shifted, leading to a process of territorialisation. Over the course of the twentieth century, the international laws and institutions governing the oceans changed dramatically, with the effect of much of the globe's maritime space being enclosed within resource exploitation jurisdictions known as EEZs, which reach out 200 miles from state shorelines.

A turning point in this shift was in 1934, with the publication of 'Report No. 8' by the US-Canadian International Fisheries Commission (Finley, 2011: 11–13). The report tracked the historical expansion of US and Canadian halibut fishing in the Pacific Northwest, beginning in Puget Sound in the 1880s, then northwards to Hecate Strait, and then on to the Gulf of Alaska. It found that Puget Sound was the most depleted fishery in the area, and the Gulf of Alaska was the least, with the recent imposition of regulations being associated with the recovery of fish stocks. The declines, it concluded, occurred because too many fish were caught before they spawned, and so fishers had to go further and further north, and catch smaller fish, in order to bring in the same weight in fish. If regulations could prevent fishers from catching smaller, immature fish, it would allow them to grow and reproduce, and ultimately profits would increase. Contrary to the wisdom of major figures in Western international law such as Grotius and Vattel, as well as many marine scientists such as Huxley, it was increasingly accepted that science had demonstrated the living resources of the sea to be exhaustible.

This made it possible to claim on behalf of science that new measures had to be taken which might upset some basic principles of international law. Initially, US fishing industries were opposed to research suggesting that they were engaged in overfishing. But as Japanese fishing expanded, and tensions rose in the Pacific, suspicions of overfishing shifted towards Japan, and US industry became potential agents of conservation (Finley, 2011: 27–34). In the early twentieth century, Japan vigorously promoted its fishing industry through subsidies and directed policy, as part of its attempt to challenge Euro-American imperialism in the Pacific. In 1935 the Japanese fishing industry became the largest in the world. Spokespeople for US fishing industries complained not only about increasing numbers of Japanese fishing boats but also about people of Japanese descent in the US fishing industry, with Japanese making up the largest ethnic group within it by 1918. The magazine *Pacific Fisherman* claimed in 1921 that if the magazine 'had not for the last twenty years vigorously opposed the introduction of Japanese in the American fisheries, this industry along the entire Pacific Coast would long ago have fallen into their hands' (Finley, 2011: 31). While Japanese industry heads were keen to set up a joint US–Japanese fishing company in Alaskan waters, their US counterparts were resolutely opposed, calling instead for the exclusion of Japanese fishing. Despite a State Department investigation in 1937 being unable to find conclusive evidence that Japanese vessels were fishing salmon off of Alaska, the issue came to a head in that year (Hollick, 1981: 22–4). A bill was introduced in the Senate declaring salmon hatched in Alaska to be US property, and the Alaska fishermen's association and the Pacific Coast union of seamen and longshoremen boycotted Japanese goods.

In addition to the growing scientific consensus around the occurrence of overfishing, attempts to distance Japanese fishing from US shores were also strengthened by the notion that fisheries science gave US fishers a special right to fish in certain waters. The US industry claimed a civilised advancement in fisheries science which made it uniquely responsible, unlike Japanese fishers, who blindly gathered up all the fish they could with their increasing industrial potential. As a *New York Times* correspondent wrote in a 1938 book, 'Just as the Nipponese military juggernaut rolls across China without regard for the amenities of civilization, so do these fishing vessels from Tokyo completely overlook conservation rules and principles in their quest for the finny wealth of the ocean' (Finley, 2011: 34). Beginning in 1937, scientists and industry lawyers began a campaign to overturn the customary three-mile rule in international maritime law, on the grounds of conservationism. Salmon lawyer Edward Allan, for example, argued in 1938 that the three-mile limit owed its existence to the false belief that marine fisheries were inexhaustible, a belief which had recently been disproven, and that the rule should be 'superseded by conceptions based on actualities' (Scheiber, 1986: 445).

Before the US entry into the World War II, this combination of nativist exclusionism and industry-led conservationism was beginning to change State Department thinking. Secretary of State Cordell Hull argued that because Americans had invested money in salmon hatcheries on US soil and had accepted conservationist restrictions on the catch, they legitimately possessed 'a special claim in the ocean areas beyond three miles where the fishery was centered'

(Finley, 2011: 34). The entry of the US into World War II in 1941 interrupted fishing activities in the Pacific and drew attention away from fishing disputes. But at the same time several federal agencies realised that the war offered a chance, as the General Land Office put it, to strike 'from our own thinking and international law the shackles of the three-mile limit for territorial waters' and adopt a line 'beyond the continental shelf' (Hollick, 1981: 33). All during the war, various agencies were busy fine-tuning a policy of extended jurisdiction, which was completed in January 1945 and signed by President Roosevelt in March. With Roosevelt's death in April, it was left for his successor, Harry Truman, to officially announce it in September.

This took the form of two presidential proclamations, known as the Truman Proclamations. One declared US jurisdiction over the subsoil and seabed of its continental shelf, while the other concerned fisheries. The latter proclamation was intended to deal with a particular contradiction: while salmon hatch in fresh water, in northern latitudes, and return to it to spawn, tuna have no homes and continuously cross and re-cross mostly warm waters (Finley, 2011: 47). This meant that while the US salmon industry would benefit from a policy protecting waters nearest to US soil, the US tuna industry needed access to a wide range of disparate waters, particularly off of Mexico, Costa Rica, and Ecuador. Rather than favour one or the other, the Truman Proclamation announced:

> The Government of the United States regards it as proper to establish conservation zones in those areas of the high seas contiguous to the coasts of the United States wherein fishing activities have been or in the future may be developed and maintained on a substantial scale. (in Hollick, 1981: 393)

Framing the fisheries proclamation as a conservation zone, explicitly aimed at protecting fisheries 'seriously exposed to unregulated exploitation and depletion' helped the US government to allay its concerns about unilaterally innovating international legal practices (Foreign Relations of the United States, 1945: 1496). These were the main substantial concerns around the policy within the government, particularly in the State Department's Office of Economic Affairs, which delayed its preparation (Hollick, 1981: 40). The fisheries proclamation had to be clearly differentiated from territorial sovereignty in order to be acceptable as a unilateral policy, and through the notion of a conservation zone it was intended to leave territorial sovereignty unchanged. The State Department Legal Advisor's Office compiled an official statement of legal justification to be circulated to other governments, explaining that 'No extension of territorial waters is embodied in the policy' (Foreign Relations of the United States, 1945: 1499).

At the same time, however, while separate from traditional international law, the proclamation still had to rest on some other universal principles. The legal memo argued, 'Equity and justice require that natural resources which have been built up by systematic conservation and self-denying restricted utilization, together with the industries based upon them, be protected and reserved from destructive exploitation by interests which have not contributed to their growth and development' (Foreign Relations of the United States, 1945: 1496). The proclamation furthermore conceded 'the right of any State to establish conservation zones off its shores' (Hollick, 1981: 393). These techniques avoided any discussion of how recent or continuously subject to modification the scientific knowledge was which made it possible to refer to a state 'developing' or contributing to the 'growth' of a fishery through 'self-denying restricted utilisation'.

When the *Saturday Evening Post* described the Truman Proclamations as 'one of the decisive acts of history, ranking with the discoveries of Columbus as a turning point in human destiny', it was surely from an exaggerated US-centred perspective (quoted in Anand, 1983: 165). Several states had already extended their jurisdictions beyond three miles, and soon after making the proclamations, the US government attempted to clarify or qualify them (Steinberg, 2001: 142). Yet in the years following the Truman Proclamations, many of the Latin American states increased their maritime jurisdictions, some of them out to 200 miles. While Latin American states had been declaring maritime zones for several decades, the extent and frequency of these declarations very markedly increased, and some of them were clearly modelled in part on the Truman Proclamations, no doubt in order to head off legal challenges from the US. The United States did challenge the Latin American declarations and argued that they, unlike the Truman Proclamations, were not defensible under international law (Finley, 2011: 130). US fishers also defied them, with one 1954 incident gaining a burst of media attention: a US whaling fleet owned by Greek magnate Aristotle Onassis sailed to Peruvian waters, violating Peru's declared jurisdiction, and was pursued at high speed by the Peruvian navy, with shots fired. A Peruvian court found that the whaling fleet had illegally caught around 3,000 whales and issued a fine of $3 million.

But by the 1970s, some kind of extended economic zone appeared inevitable. Moreover, with Congress's passage of the 1976 Magnuson-Stevens Act, the US unilaterally established its own 200-mile fishing zone (Kraska, 2011: 142). In a similar way to the 1945 Truman Proclamation, this contributed to the growing trend towards international adoption of zones of this size and undermined any US arguments against it. In order to ensure that freedom of navigation within this economic zone would be maintained, then, the US signaled agreement. This led eventually to the establishment of the EEZ, enshrined in the UNCLOS III agreement concluded in 1982. As the conversation had gone in 1971 between US President Richard Nixon and National Security Advisor Henry Kissinger:

> **Nixon:** I don't give a damn about the fisheries anyway. Let everybody have 200 miles to fish. They're all poverty-stricken down there [in Brazil] anyway.
> **Kissinger:** If we dig in on the fisheries, we'll lose on navigation—
> **Nixon:** Navigation we want. Let them fish if they want. That's my view.
> **Kissinger:** Well, that's my recommendation, Mr. President.
> (quoted in Kraska, 2011: 140)

Conclusion

Taking the ocean seriously is essential for any discipline whose object of analysis is coextensive with the 'world' or the 'international'. It would be a grave mistake to unthinkingly assume that, if people do not make homelands out of the sea in the same way they make them out of the land, the ocean is merely a periphery to international relations. But just as international relations should be studied beyond human societies' terrestrial homelands, taking the ocean seriously should lead us beyond what happens on and in the ocean. If the ocean is not simply empty space, as far as IR is concerned, one reason for this is that oceans make possible certain kinds of relations between humans and other life forms which have important implications for international relations. As we have seen, for instance, the way in which states have projected their territorial jurisdictions beyond land, or not, has depended heavily on societies' relationships with fish. The complexities of overfishing processes, and the way in which these complexities are refracted through political, scientific, and legal debates in particular places and times, have been highlighted here as central for understanding the origins of the EEZ and the territorialisation of the ocean.

In engaging with the key themes of the volume, this chapter has shown the interconnectedness between two roles the sea has often played in relevant literature, that of a wilderness to be controlled and that of resources to be extracted. For example, those who constructed a discipline of fisheries science, in order to increase and maintain high landing volumes, had to make scientifically knowable and measurable the life cycles and movements of different species of fish which were otherwise known only to fishing communities, or were simply mysterious. What was at stake here was not only the quantity of fish landed but also the ability to control the supply of fish and to create knowledge about fish which would be useful to states and large industries.

While fish remain a crucial food source in many parts of the world, the use of fishing not only to feed people and provide livelihoods but also simply to control space is arguably as prevalent as ever. Despite the relatively small size of the UK fishing industry, for example, having contributed £784m to 2018 GDP compared to financial services' £132bn, the promise of taking back control of UK fishing waters was potent rhetoric in favour of the Brexit campaign (Morris, 2020). Political principles of 'fairness' and 'sovereignty' here matter far more than rational economic logics. China's fishing fleet, in a somewhat parallel way, does not only catch fish but also reinforces China's exaggerated claims to maritime control, as mentioned previously. These examples point to a recurring imperative to dominate maritime space going beyond the instrumental need for any particular set of resources.

But while this chapter has focused on the example of fish, one might examine all kinds of human–environment interactions in a similar way. While a state seeking to maximise its fishing yields, by either imposing a territorial jurisdiction on the sea or not, runs the risk of misjudging its environment and devastating a future food supply, territoriality on land equally involves human–environment relationships which are often neglected (Mukerji, 1997). Whether on land or not, fixing a boundary along a line of latitude, for example, assumes a certain knowledge of the curvature of the earth and the distance from the equator, as well as the ability to manipulate objects so that the boundary remains fixed. During the Renaissance and the Enlightenment, land drainage projects involving a struggle between the rational mind and chaotic wetlands allowed rulers, as Frederick the Great of Prussia put it, to 'conquer' new territory (Yao, 2019).

The absence of the ocean in our historical accounts of international relations should thus draw attention to the absence of the environment in general. Environmental history does not exist only to help broaden the horizons of climate scientists and activists and to provide greater perspective on the social and political effects of ongoing environmental degradation, although these are of course important activities. As this chapter has attempted to show, it also has much to offer to IR as history in its own right, as the once rigid distinction between natural and human history continues to break down.

Notes

1 Quoted in Hunt, 2017: 45.
2 I argue this point in more detail in Goettlich (2019).
3 See, for example, Deudney (1999); for an overview of these literatures, see Fagan (2017).
4 Tisquantum, better known to some as Squanto, was sold in Spain, and then through events of which there is little record, somehow made his way back to New England, via England and Newfoundland (Adolf, 1964: 248).
5 The historical claims China uses here have been subject to serious critique, although this does not contradict the argument that fishing is often important for maritime empires (Haydon, 2017).

References

Adolf, L. (1964). 'Squanto's role in Pilgrim diplomacy', *Ethnohistory* 11(3): 247–61.

Alder, J., and U. R. Sumaila (2004). 'Western Africa: a fish basket of Europe past and present', *Journal of Environment and Development* 13(2): 156–78.

Anand, R. P. (1983). *The Origin and Development of the Law of the Sea*, The Hague: Martinus Nijhoff.

Antonova, A. (2016). 'The rhetoric of "responsible fishing": notions of human rights and sustainability in the European Union's bilateral fishing agreements with developing states', *Marine Policy* 70: 77–84.

Aswani, S. (2005). 'Customary sea tenure in Oceania as a case of rights-based fishery management: does it work?' *Reviews in Fish Biology and Fisheries* 15: 285–307.

Atta-Mills, J., J. Alder and U. R. Sumaila (2004). 'The decline of a regional fishing nation: the case of Ghana and West Africa', *Natural Resources Forum* 28: 13–21.

Barkawi, T. (2009). 'Grounded', *Review of International Studies* 35(4): 859–70.

Belhabib, D., U. Rashid Sumaila, V. W. Y. Lam, D. Zeller, P. Le Pillon, E. Abou Kane and D. Pauly (2015). 'Euros vs. Yuan: comparing European and Chinese fishing access in West Africa', *PLOS ONE* 10(3): 1–22.

Bueger, C. (2013). 'Practice, pirates and coast guards: the grand narrative of Somali piracy', *Third World Quarterly* 34(10): 1811–27.

de Carvalho, B., and H. Leira (2022). 'Introduction: staring at the sea', in B. de Carvalho and H. Leira, eds, *The Sea and International Relations*, Manchester: Manchester University Press.

Chakrabarty, D. (2009a). *Climate of History in a Planetary Age*, Chicago: Chicago University Press.

Chakrabarty, D. (2009b). 'The climate of history: four theses', *Critical Inquiry* 35(2): 197–222.

Chasek, P., D. Downie and J. Brown (2018). *Global Environmental Politics*, New York: Routledge.

Cronon, W. (2003). *Changes in the Land: Indians, Colonists, and the Ecology of New England*, New York: Hill and Wang.

Dagne, T. (2009). 'Somalia: prospects for a lasting peace', *Mediterranean Quarterly* 20(2): 95–112.

Dalby, S. (2002). *Environmental Security*, Minneapolis, MN: University of Minnesota Press.

Davey, J. (2018). 'Introduction', in J. Davey, ed., *Tudor and Stuart Seafarers: The Emergence of a Maritime Nation, 1485–1707*, London: Bloomsbury Publishing, 8–19.

Deudney, D. (1999). 'Bringing nature back in: geopolitical theory from the Greeks to the global era', in D. Deudney and R. Matthew, eds, *Contested Grounds: Security and Conflict in the New Environmental Politics*, Albany, NY: SUNY Press, 25–57.

Durrenberger, E. P., and G. Palsson (1987). 'Ownership at sea: fishing territories and access to sea resources', *American Ethnologist* 14(3): 508–22.

Edmond, G. (1995). 'The freedom of histories: reassessing Grotius on the sea', *Law/Text/Culture* 2: 179–217.

Elden, S. (2010). 'Land, terrain, territory', *Progress in Human Geography* 34(6): 799–817.

Fagan, M. (2017). 'Security in the anthropocene: environment, ecology, escape', *European Journal of International Relations* 23(2): 292–314.

Fenn, P. T. Jr. (1926). *The Origin of the Right of Fishery in Territorial Waters*, Cambridge, MA: Harvard University Press.

Finley, C. (2011). *All the Fish in the Sea: Maximum Sustainable Yield and the Failure of Fisheries Management*, Chicago: Chicago University Press.

Fishel, S., A. Burke, A. Mitchell, S. Dalby and D. Levine (2016). 'Planet politics: a manifesto from the end of IR', *Millennium: Journal of International Studies* 44(3): 499–523.

Fletcher, E. (1933). 'John Selden (author of *Mare Clausum*) and his contribution to international law', *Transactions of the Grotius Society* 19: 1–12.

Foreign Relations of the United States: Diplomatic Papers (1945). 'Explanatory statement on the protection and conservation of coastal fisheries', *Foreign Relations of the United States: Diplomatic Papers, 1945, vol. II*.

Gagern, A., and J. van den Bergh (2012). 'A critical review of fishing agreements with tropical developing countries', *Marine Policy* 38: 375–86.

Gao, Z., and B. B. Jia (2013). 'The nine-dash line in the South China Sea: history, status, and implications', *American Journal of International Law* 107(1): 98–123.

Goettlich, K. (2019). 'The rise of linear borders in world politics', *European Journal of International Relations* 25(1): 203–28.

Grotius, H. (2004). *The Free Sea*, D. Armitage, ed., Indianapolis, IN: Liberty Fund.

Hansen, S. J. (2011). 'Debunking the piracy myth: how illegal fishing really interacts with piracy in East Africa', *RUSI Journal* 156(6): 26–31.

Hiller, J. K. (1996). 'The Newfoundland fisheries issue in Anglo-French treaties, 1713–1904', *The Journal of Imperial and Commonwealth History* 24(1): 1–23.

Hoffmann, R. (1996). 'Economic development and aquatic ecosystems in medieval Europe', *The American Historical Review* 101(3): 631–69.

Hollick, A. (1981). *U.S. Foreign Policy and the Law of the Sea*, Princeton, NJ: Princeton University Press.

Hubbard, J. (2011). *A Science on the Scales: The Rise of Canadian Atlantic Fisheries Biology, 1898–1939*, Toronto: University of Toronto Press.

Hubbard, J. (2014). 'In the wake of politics: the political and economic construction of fisheries biology, 1860–1970', *Isis* 105(2): 364–78.

Hunt, K. (2017). *Herring: A Global History*, London: Reaktion.

Iheduru, O. C. (1995). 'The political economy of Euro-African fishing agreements', *The Journal of Developing Areas* 30(1): 63–90.

Kaczynski, V., and D. Fluharty (2002). 'European policies in West Africa: who benefits from fisheries agreements?' *Marine Policy* 26: 75–93.

Kamola, I. (2018). 'Pirate capitalism, or the primitive accumulation of capital itself', *Millennium: Journal of International Studies* 47(1): 3–24.

Kraska, J. (2011). *Maritime Power and the Law of the Sea: Expeditionary Operations in World Politics*, Oxford: Oxford University Press.

Levi, L., ed. (1866). *Annals of British Legislation: Being a Digest of the Parliamentary Blue Books*, vol. 3, London: Smith, Elder, and Co.

Mearsheimer, J. (2001). *The Tragedy of Great Power Politics*, New York: W.W. Norton.

Meltzer, E. (1994). 'Global overview of straddling and highly migratory fish stocks: the nonsustainable nature of high seas fisheries', *Ocean Development and International Law* 25: 255–344.

Morris, C. (2020). 'Fishing: why is fishing important in Brexit trade talks?' *BBC News*, 16 June 2020, www.bbc.com/news/46401558 (accessed 14 June 2021).

Mukerji, C. (1997). *Territorial Ambitions and the Gardens of Versailles*, Cambridge: Cambridge University Press.

O'Neill, K. (2009). *The Environment and International Relations*, Cambridge: Cambridge University Press.

Onley, J. (2009). 'Britain and the Gulf Shaikhdoms, 1820–1971: the politics of protection', Occasional Paper No. 4, Center for International and Regional Studies, Georgetown University School of Foreign Service in Qatar.

Pauly, D., V. Christensen, S. Guénette, T. J. Pitcher, U. R. Sumaila, C. J. Walters, R. Watson and D. Zeller (2002). 'Towards sustainability in world fisheries', *Nature* 418: 689–95.

Samatar, A. I., M. Lindberg and B. Mahayni (2010). 'The dialectics of piracy in Somalia: the rich versus the poor', *Third World Quarterly* 31(8): 1377–94.

Scheiber, H. (1986). 'Pacific Ocean resources, science, and law of the sea: Wilbert M. Chapman and the Pacific Fisheries, 1945–70', *Ecology Law Quarterly* 13(3): 381–534.

Sevastopulo, D., and K. Hille (2019). 'US warns China on aggressive acts by fishing boats and coast guard', *Financial Times*, 28 April 2019, www.ft.com/content/ab4b1602-696a-11e9-80c7-60ee53e6681d (accessed 14 June 2021).

Smed, J., and J. Ramster (2002). 'Overfishing, science, and politics: the background in the 1890s to the foundation of the International Council for the Exploration of the Sea', *ICES Marine Symposia* 215: 13–21.

Sokolsky, M. (2015). 'Fishing for empire: settlement and maritime conflict in the Russian Far East', *Arcadia* 20, https://arcadia.ub.uni-muenchen.de/arcadia/article/view/92 (accessed 10 February 2022).

Steinberg, P. (2001). *The Social Construction of the Ocean*, Cambridge: Cambridge University Press.

de Vattel, E. (2008). *The Law of Nations*, ed. B. Kapossy and R. Whatmore. Indianapolis, IN: Liberty Fund.

Waltz, K. (1979). *Theory of International Politics*, Reading, MA: Addison-Wesley.

Welwood, W. (2004). 'Of the community and propriety of the seas', in D. Armitage, ed., *The Free Sea*, Indianapolis, IN: Liberty Fund, 63–74.

White, P. (2003). *Thomas Huxley: Making the 'Man of Science'*, Cambridge: Cambridge University Press:

Wilson, C. H. (1967). 'Trade, society and the state', in E. E. Rich and C. H. Wilson, eds, *The Cambridge Economic History of Europe, Volume IV*, Cambridge: Cambridge University Press, 487–576.

Yao, J. (2019). ' "Conquest from barbarism": The Danube Commission, international order and the control of nature as a Standard of Civilization', *European Journal of International Relations* 25(2): 335–59.

Ziskind, J. (1973). 'International law and ancient sources: Grotius and Selden', *The Review of Politics* 35(4): 537–59.

Conclusion: international terraqueous relations

Xavier Guillaume and Julia Costa López

International Relations (IR) has long ignored the sea. As the editors of this volume emphasise, the relative ahistoricism of the discipline, along with a focus on ostensibly 'landed' concepts, have until very recently led to a 'trend of oceanic amnesia' (de Carvalho and Leira, this volume), against which *The Sea and International Relations* militates. And yet, as the contributions in this volume bring to the fore, this was only ever a fictitious absence. For the sea was an ever-present silent counterpart to most of IR's conceptual apparatus and subject matter. And therein lies what in our view is the core contribution of this volume. Bringing the sea into IR is not (only) a matter of paying attention to something hitherto ignored, but also forces us to rethink IR's core histories and frameworks. Indeed, taken together the chapters in this volume clearly highlight the necessity to move away from facile dichotomies, predominant in IR, whereby land and sea would be the two faces of the international coin: the land is a governable, delineated, and sovereign space whereas the sea is its ungovernable, unlimited, and anarchic counterpart.

Against this binary understanding, *The Sea and International Relations* opens up an alternative and productive way of thinking about the sea in IR: terraqueous relations (see Bashford, 2017). Artificially separating processes as pertaining to either land or water may miss the point of how not only both spaces are connected but how their interconnections are the condition of possibility, materially and symbolically, of the international itself. As Colás' chapter points out, it is necessary to reflect on the close material connections between land and water to comprehend the different temporalities and spatialities that constitute the international. In this conclusion, we bring together the different terraqueous temporalities and spatialities highlighted by the chapters in the volume to consider what an international terraqueous relations analytics may look like and how it may help overcome the limitations in the current IR literature.

The various contributions go a long way in demonstrating the value of a terraqueous perspective of international relations. The historical lineages of exploration, exploitation, and conflicts over the seas, and especially the Atlantic, are at the centre of Shirk's chapter on 'the line' and de Carvalho and Leira's on Protestant privateering. What these chapters clearly highlight is how the practice of the sea has come to define how not only the sea became a specific political space, a terraqueous space across the Atlantic shores, but also how the economic ventures through which Protestant privateers came to wreak havoc on Catholic fleets also 'came to define the main structures of the colonial world' that emerged in the seventeenth century onwards. These temporal lineages are important reminders that while we can easily agree that the sea is a politically constructed space, the central question remains how it is so, and what is thereby constructed.

Many chapters thus reconstruct how apparently technical, scientific, or legal practices dealing with the sea have to be read politically in how they construct specific spaces which enact different outcomes. Braarud, for instance, discusses the tensions between legally designating oceans a 'common heritage of mankind [sic]' while at the same time manoeuvring to exploit deep seabeds. Dickson, for her part, demonstrates how it is insufficient to concentrate on how a space-transformed border such as the Mediterranean is legally ordered and enacted; we also need to consider 'how seafaring practices are changing a way of ordering' by 'laying down the pathway for manipulations and extenuations of its very own structures'. More fundamentally, as highlighted by Goettlich in this volume, the centrality of the political in determining what oceans are to us uncovers the blind spots of our conceptualisations about the international, as they are reflective of this arbitrary and artificial separation of human history from natural history; the environment remaining an afterthought in our efforts to think the international, rather than one of its central features and actant. This important question invites further reflections as to how the political spatiality of the sea is constructed. Leira and de Carvalho's chapter on the gendered and racialised constructions of the sea reminds us

246 The Sea and International Relations

about the different ways by which the feminine and masculine, the racialised self and other, the gendered roles of the sea have played out in fixing the political boundaries of the sea, and with it, those of land. In the same vein, Mälksoo and Simonds in their respective chapters highlight further venues to reflect on how routinised and ritualised practices concerned with the sea help construct collective identities and spaces.

The Sea and International Relations thus covers important ground in 'bringing the sea' to IR. Every chapter offers an invitation to explore further historical lineages, constructions of collective identities, and spaces. We would like to take on this invitation to explore further research avenues on what International Terraqueous Relations may look like by highlighting three issues that, we believe, are still in need of further elaboration. First, while the call to 'bring in the sea' is an important one, there is a lingering question as to the manner by which this reintegration should be made (see Goettlich in this volume). The dominant vision about the sea in the field is rather detached from considerations about the Western-centric imaginaries that are connecting the sea as a space to the processes which have historically constituted international *relations*. As we first discuss, an invitation to bring the sea into IR should also consider that until now the sea has been largely connected, whether explicitly or implicitly, to a Western *thalassodicy*, a portrayal of the sea by which the West, and especially an Anglo-American West, rationalises and legitimises its moral, political, and military power over others. Thus, the question becomes how to move beyond such *thalassodicy*.

Second, bringing in the sea to IR should also draw our attention to the multiple forms of international dynamics that are fostered through terraqueous relations. While most analyses in this volume focus on meso- or macro-level dynamics, what is often termed as 'the experience of the sea' is also an invitation to concentrate on micro-level practices, or more broadly on the sea as everyday terraqueous relations. Inspired notably by feminist scholarship and other approaches to quotidian international relations, bringing back the sea should also be concerned with the racialised and gendered migrant, the maritime economic proletariat, the ecological exploitation and spoliation of the seas and its consequences on the daily lives on entire populations, the place and role of the security professionals of the sea, or the individuals and groups bound by the sea at the source of different social movements that have come to define the international.

Finally, the call to bring the sea into IR should also consider that the sea is not a homogeneous and self-evident space. The 'sea', a term that is used interchangeably to speak about oceans and seas, as with any natural space from mountains to the idea of landscape, is a social construct as well. As we argue, to engage with international terraqueous relations requires designing analytically precise tools recognising the differences, as well as the similarities, between different terraqueous spaces such as oceans, seas, and lakes. Doing so, we think, offers a vantage point from which to examine how social imaginaries, practices, and ecosystems interact.

The Western *thalassodicy*: the sea as a Western imaginary of geopolitical power

While the current volume very clearly highlights how dichotomies are by and large superfluous when discussing the sea, it is worth still reflecting on the predominance of such imaginaries in current IR literature. We believe that this predominance is related to what we term a Western *thalassodicy*, a conceptualisation of the sea by which the West, and especially an Anglo-American West, justifies and legitimises its moral, political, and military power over others. *Thalassodicy* is a neologism describing how modern and contemporary Western thinkers have come to rationalise and teleologically justify this Western dominance by contrasting a specific 'essence' behind countries which turn to the sea, and those which are anchored to the land. This calls attention to the sea as a *geo*political notion, a specific geo-graphing, to borrow from Gearóid Ó Tuathail, which is born out of the 'centralization and imperialist expansion of the modern European state system across the globe from the sixteenth century onward' (1996: 8). Geopolitics as a discourse is precisely that ordering of spaces (land and sea) into 'particularistic regimes of nationalistic, ideological, racial, and civilizational truth' (Ó Tuathail, 1996: 12) justifying the power of some over others, or a certain global hierarchy. In this sense, it is no surprise that the shift towards the study of the sea in the early twentieth century in disciplines such as history and geography was

closely connected to the scholarly tradition of geopolitics, with figures such as Mackinder, Haushofer, or Carl Schmitt figuring prominently (Bashford, 2017: 256). A terraqueous analytics of the sea
thus serves to highlight both how the notion of the sea that stems
out of this tradition has been implicated in the construction of
Western dominance, and how IR continues to reproduce this by
understanding the sea within a land/sea binary.

Andrew Lambert's recent *Seapower States* (2018) serves as an
illustration of how a particular dichotomisation of land and sea
serves to rationalise Western dominance. *Seapower States* describes
how some states made the 'cultural choice' to become seapowers, a
choice that 'was facilitated by inclusive politics, a democracy or an
oligarchic republic' (Lambert, 2018: 323), not because they were
or could become naval powers but because as seapower states –
Athens, Carthage, Venice, the Dutch Republic, Britain and ultimately, and teleologically, in Lambert's account, the United States
of America – they had to. Enemies of seapowers – including the
Ancient Mesopotamian kingdoms, Persia, the Roman Republic,
Ottoman Turkey, France, Russia/U.S.S.R., Germany, and China –
'fear the inclusive, progressive ideologies of seapower, using armed
force, on land and sea, to destroy the cultural challenge' (Lambert,
2018: 324).[1] Lambert's is a *thalassodicy* of Anglo-Americanism
through which the sea is constructed as a marker of civilisation,
not unlike Orientalist accounts or earliest geopolitical accounts like
Alfred T. Mahan's. According to Lambert, Western *seapowers* were
and are relatively stronger than their absolutist enemies because
of their moral, political, technological, economic, and military
superiority. Western seapowers 'did more to advance trade, knowledge and political inclusion than their landed peers: they shaped
the global economy and the liberal values that define the contemporary Western world', values without which 'the world would
be a darker place, lacking the cultural diversity and exchange that
sustains creativity' (Lambert, 2018: 226). The land/sea dichotomy
is thus constituted through and superposed onto a variety of
binaries – static/creative, progressive/conservative, etc. – which not
only create a historical metanarrative of progress, but also serve to
morally justify Western dominance.

Indeed, in these accounts the sea appears as a space of exclusively positive connections, where seapowers can foster 'cultural

diversity', 'creativity', or 'political inclusion' (Lambert, 2018). At the same time, however, crucial dynamics of power, violence, and exploitation – such as the place of the Middle Passage as a central part of the Dutch and British political economies or how slavery was a central institution of the Athenian political system – completely disappear from the story. And yet, a terraqueous analytics cautions us from addressing these omissions by simply adding them to the story, for in doing so, and in leaving the constitutive conceptual binaries intact, we risk reinscribing them into a Eurocentric perspective that reproduces the 'assumption of European centrality in the human past and present' (Barkawi and Laffey, 2006: 331). Indeed, as Çapan has noted (2020), the predominant understanding of Eurocentrism as absence of the non-European, and thus the presentation of the solution as its reincorporation, risks leaving the core Eurocentric interpretive schemes untouched, thus reinscribing them into the research. The critique has been levied also on important works on the sea. As seminal a work as Gilroy's *Black Atlantic* can be read through the lens of its omissions and the assumptions it leaves intact: despite recovering previously ignored Atlantic dynamics of power, exploitation, and resistance, it does so by focusing chiefly on the Anglophone and American parts of the Atlantic. The resulting image, Piot (2001: 156) claims, is that 'Africa has played little role in the development of black Atlantic cultural production, other than as provider of raw materials – bodies and cultural templates/ origins – that were then processed or elaborated upon by the improvisational cultures of the Americas.' And it doing so it risks reproducing the very interpretive schemes that it sought to challenge (see also Shilliam, 2015).

If Western *thalassodicy* is such a central element of existing approaches to the sea, the question then becomes how IR can move past it. What are the counter-stories and counternarratives that IR can tell about the sea – counter-stories and counternarratives that would move away from the various forms of Western *thalassodicy*? A possibility is to offer connected histories of the international, forms of historical narratives and research that historians have produced for more than three decades about the sea and its place in the shaping of early modern and modern 'international relations' (for a recent assessment in IR, see Çapan, 2020). Sanjay Subrahmanyam, for instance, in his seminal article 'Connected histories' (1997),

invited us to write histories that move away from a largely teleological and particular European trajectory from ancient Greece to
'modernity', to concentrate not only on the 'many different sources
and roots, and – inevitably – many different forms and meanings'
of the latter, but also how 'modernity' was a 'more-or-less global
shift' (Subrahmanyam, 1997: 737). What makes these sources and
roots protean often is their relations to the sea. French historian
Romain Bertrand (2011: 84–93), for instance, describes the similarities between Insulindian and European nautical imaginaries in
the seventeenth century, and how Insulindian nautical technologies,
from ships to cartographies, were much more developed than those
of the Europeans, who ended up relying heavily on Insulindians'
knowledge of the sea. A connected approach to the sea thus offers
a distinct analytics that goes beyond the focus on substantivist
entities to centre processes, connections, and flows.

And yet, challenging Western-centric understandings of the sea
requires thinking beyond mere connections and considering also
disconnection. For an exclusive focus on connections risks reproducing familiar teleological narratives about modern globalisation as an increasingly intensifying process of connection, enabled
by the sea (see e.g. Buzan and Lawson, 2014; for a critique, see
Costa López, forthcoming). A relational analytics that considers
connection and disconnection alongside one another, conversely,
brings to light the effects of the termination of connection, and the
hierarchies and power dynamics in and of the sea. From a macro
perspective, for example, an exclusive focus on connection in
Atlantic history would not be able to capture nineteenth-century
dynamics: a shift towards the Pacific and Indian Oceans, the end
of the slave trade and the independence of the Spanish colonies all
mean that to a large extent 'the two sides of the Atlantic drifted
apart in reality and in people's minds' (Osterhammel, 2014: 100).
Similarly – and to a certain extent relatedly – the building of the
Suez Canal in 1859 cannot only be understood as both contributing to increasingly dense connectivity, but also causing a variety
of dynamics of disconnection. In terms of trading routes, it led to
the decline of existing sea routes through the Cape of Good Hope,
and to alterations in Asian trading routes. Beyond this, at the same
time as it fostered the production of steamships – which were able
to navigate the canal on their own – it led to the decline of sailships

which, needing to be dragged for the entire canal, became more expensive (Fletcher, 1958). And yet very recently the proliferation of piracy off the Horn of Africa has led a large number of companies to reroute again via the Cape of Good Hope, reducing traffic through the Suez Canal (Sullivan, 2010).

A terraqueous analytics of (dis)connection can thus help challenge a Western *thalassodicy* by capturing the fundamental role of the transformation of connections. In the East and South China Seas, the Insulindian seas and the western Indian Ocean, cultural and commercial connections and networks that existed long before European incursions persisted beyond their arrival and integrated the Europeans in them (Bertrand, 2011). The Atlantic Ocean never forgot the triangular trade behind the Middle Passage. Yet, beyond all the sufferings, the sea nonetheless created one of the most potent counternarratives to the Western *thalassodicy*, that of the black Atlantic: that 'fractal structure of transcultural, international formation', which 'transcend[s] both the structures of the nation state and the constraints of ethnicity and national particularity' (Gilroy, 1993: 4, 19). Paul Gilroy's invitation to a 'theorisation of creolisation, métissage, mestizaje, and hybridity' is a powerful reminder of how the human, cultural, intellectual, and economic circulations across the Atlantic have fostered the possibility to politically think differently about how ethnicity and nationality work in the connected spaces of the Caribbean, American, Africa, and Europe beyond convenient and comforting dichotomies (Gilroy, 1993: 2, 11–19; Getachew, 2019).

The experience of the sea: everyday terraqueous relations

A second element in need of further attention is the level and types of processes that are studied under the banner of 'the sea in IR'. Terraqueous relations are plural in their dynamics, but for the most part the literature in IR seems to concentrate on terraqueous realities that pertain to states – from naval power to piracy – or to disembodied and fetishised economic flows or legal questions. The current tragic attempts by migrants to traverse the Mediterranean Sea at the risk of their lives and at the mercy of both criminal organisations and state instrumentalisation are a tragic reminder that terraqueous relations are an ongoing, quotidian reality for those who live by/from/through the sea. What we see missing is a larger discussion of the everydayness of terraqueous relations. The everyday is a text that is relevant to illuminate central practices at the heart of the production of 'international' representations and their effects, which are as concerned with the reproduction and contestation of relations of domination individuals and groups face as they are with the reproduction of military, legal, and economic state relations (see Guillaume, 2011). The everydayness of terraqueous relations is concerned with the racialised and gendered migrant, the maritime economic proletariat, the ecological exploitation and spoliation of the seas and its consequences on the daily lives on entire populations, the security professionals of the sea, or the individuals and groups bound by the sea at the heart of social movements that gave to what we call in modern times the international its first cosmopolitan textures.

Drawing inspiration from feminist scholarship in IR, such as Cynthia Enloe's (2000 [1989]: 195–201) famous observation that the personal is international and the international is personal, international terraqueous relations would gain from the examples offered by historians in drawing connections between everyday processes and the emergence, reproduction, or transformation of international processes. While the realm of the everyday is not devoid of a certain romanticism and requires a rigorous analytical discussion every time it is mobilised (see Guillaume and Huysmans, 2019), the field of IR can easily identify everyday figures of the experience of the sea that are relevant even to its most classical concerns such as security. For instance, a central figure in nuclear deterrence are sonar operators serving in attack or ballistic missile submarines (see Guillaume and Grayson, 2021). Across the world's oceans, attack submarines seek out the locations of ballistic and cruise missile submarines that, in turn, seek to evade them. Since the 1950s, sonar operators are crucial figures in contemporary navies; they 'are looked upon highly, and are constantly depended on for the ship's and crew's safety' (Hedrick, 2015). Furthermore, sonar operators are central to the larger configuration of individuals, devices, protocols, practices, doctrines, and institutions that constitute the architecture of nuclear deterrence. However, sonar operators, as with many everyday figures of the sea, are largely left out of any research on the everyday of the international. There are plenty of figures and their related quotidian communities to research synchronically and diachronically; communities that have a distinguishing feature in being terraqueous. They are communities of exchange and commerce (e.g. merchants, diasporas), communities of labour (e.g., contracted sailors or fishermen), communities of security (e.g., naval officers and sailors), and communities of mobility (e.g. migrants).

Another dimension of everyday terraqueous relations that could benefit from further attention concerns a central object and space of our globalised worlds, an obvious yet 'largely neglected figure in its own right': the ship (see Hasty and Peters, 2012: 660, passim). As noted in the early 1990s by Paul Gilroy (1993: 4), the ship is 'a living, micro-cultural, micro-political system in motion'. It is also by and large one of the most impactful objects ever made by humankind, playing a major role in shaping the human world from its earliest age to now. It is impossible to understand our current worlds as they have been shaped by global capitalism, colonialism, and imperialism without the largely invisible role of ships. As noted by David Lambert, Luciana Martins, and Miles Ogborn (2006: 487), ships present in themselves different spaces, which means different forms of everyday terraqueous relations. Whether we are speaking of aircraft carriers, high ocean fishing boats, migrant rafts, pirate ships, nuclear submarines, oil tankers, or container ships, the type of communities on board, how they interact within but also sometimes between these different spaces, and how these communities are embedded in terraqueous spaces such as ports, are an important dimension of how the international is personal, and the personal international. This is crucial as 'ways of living with the sea are as differentiated by gender, class, race and sexuality, as the production of space, place and landscape' (Lambert et al., 2006: 487).

The analytical importance of ships is not limited to the myriad ways of living by or in the sea. As the previous section emphasised, a terraqueous perspective calls for an analytics of (dis)connection. This is not just a matter of abstract concepts: the everydayness of international terraqueous relations is also the actual material and physical relationalities the ships are performing by connecting two points in space (see Law and Mol, 2001: 611–13). Following Bruno Latour, John Law and Annemarie Mol consider that the ship both constitutes the material determination of networks – trading lines which are part of the networks of global capitalism, fishing routes which are part of the networks of the global political economy of food, cruise lines which are part of the networks of the global political economy of tourism, etc. – but also performs the relationalities that determine what that network is. It does so in two different ways, as it articulates two 'topological systems'. On the one hand, the ship is a mobile object that is immutable as a space. The ship, which is all its parts – the financing and work to construct it and employ it, the crew to manoeuvre it, etc. – is subjected to its own determinations as an 'immutable mobile' which, while it has changed significantly across centuries, remains the same object. On the other hand, as a mobile object, it connects physically different Euclidean points on the planet, an Euclidean space, and in doing so it performs a network space, e.g. an empire, global capitalism, global tourism, etc., which is immutable to the extent that the ship plays a specific role in that network space – it transports slaves, commodities, and so on, it unleashes violence, etc.

In other words, the broad dynamics and processes that IR is interested in – power, domination, oppression, emancipation, etc. – exist only insofar as they are every day performed both onboard and through it (see Gilroy, 1993; Linebaugh and Rediker, 2000). The consequence of this double determination is that to understand changes in these dynamics and networked spaces – e.g. to understand change in imperial governance or the expansion of trade networks – it is not sufficient to change how the different elements of that space relate in Euclidean terms. It is not sufficient, for example, to merely focus on port cities, or capitals, or mines, as that would mean giving a specific point independence to change the ways in which it is shaped by the network space. Rather, understanding change in a connected terraqueous network also requires tracing the ways in which the everyday relationalities themselves change and determine the space. If IR is to understand core processes of change in international terraqueous relations, therefore, it is essential to look at how the crucial vectors of these relations – such as ships – work both onboard *and* as vectors across the Euclidean space.

Oceans, seas, and lakes: materialising the multitude of terraqueous relations

A final issue which is not necessarily fully covered by the current volume, and would be worth paying attention to in IR, is the general lack of analytical and conceptual differentiation between the different terraqueous relations that can emerge from their taking place within or across different types of basin – such as oceans, seas, rivers, and lakes – and in specific basins among these. The use of the term 'sea', for example, still tends to encompass both oceanic or sea basins (Lewis, 1999: 208). On the one hand, the use of the sea as singular has been key in helping think through the distinctive materiality of water and its consequences for social activity. At the same time, however, this monolithic understanding of water basins remains tied to the sea/land binary, this time through material associations of solidity/fluidity and everything that ensues (Steinberg and Peters, 2015). Against this, in this conclusion we raise the epistemological question of the compatibility and transferability of analytical models across types of water basins (Lambert et al., 2006: 482). For instance, is, and if so how and to what extent, Fernand Braudel's 1949 seminal *La Méditerranée et le monde méditerranéen à l'époque de Philippe II* (2017 [1979]) and its *longue durée* perspective adaptable to a context such as the Atlantic Ocean as exemplified in Gilroy's 'long contemporaneity' in the Black Atlantic (see Baucom, 2001: 10–11)? The notion of terraqueous international relations invites us to not treat 'the sea' monolithically, and to think about how we come to conceptualise and empirically treat distinct water basins. This not only remains a challenge in IR, but also offers an opportunity for novel theorisation.

A first way of unpacking the multiplicity of water basins and their individual specificity would draw attention to their socially constructed nature. As some chapters in this volume emphasise, imaginaries play a crucial role in the definition of specific water basins. Historically, for instance, the division into oceans, which are now considered to be three or four depending on whether or not you count the Arctic Ocean, is a relatively recent development (see Lewis, 1999). Whereas the oceans were originally in European mythological and cartographical imaginations conceptualised as a single entity, they gradually became 'discrete units of sea space' (Lewis, 1999: 199), which are now internationally and officially delineated by the International Hydrographic Organization. And yet, the delineation of water basins is far from a purely technical exercise, but has important political dimensions. For example, a crucial part of the mobilisation strategy in the environmentalist movement, the UN, and scientific circles (Sivasundaram et al., 2018: 1), has been to once again speak of 'the world ocean', as a move that serves to highlight the essential role of water basins for planetary life. This perspective also brings to the fore the fundamental question of naming. Historically, the naming of a sea has been a crucial step in the claiming of control over it. Not only does it help in tracing the boundaries of a particular water basin, and thus constitute them, but as the dispute over the naming of the Sea of Japan/East Sea shows, also constitutes the basis for contestation between different political imaginations (Dudden, 2018).

Individual water basins, however, are constituted by more than naming practices and imaginaries. Indeed, specific sets of social practices and interactions are crucial in constituting water basins into distinct social spaces. Fishing locations, trading routes, seabed exploitation, or cruiseship tours all create distinct spatialities and geographies that prevent us from thinking about the sea in a unitary way. Historically, for example, it is specific practical configurations that allow us to speak of 'the Atlantic' as a relevant object of analysis. Triangular trade, the violence of the slave ships, or plantations all constituted the Atlantic into a particular social world that connected peoples thousands of kilometres apart. And yet, as a distinct space, one cannot but note that this historical, practically constituted Atlantic of triangular trade and connections was not perfectly coterminous with other Atlantics with which it coexisted. At a basic level, the spatiality of the Black Atlantic was significantly different from a larger, 'purely' hydrological Atlantic (more on this below), but it also differed substantively from the 'distinctly white and racially charged' Atlantic of early twentieth-century geopolitics (Armitage, 2018: 92; see also Bell, 2020).

In placing the constitution of distinct spatialities front and centre, a terraqueous perspective shows the fundamental limitations of the land/sea binary. For the understanding of a specific sea or lake as a social space needs necessarily to include the (landed) shore regions which border the water basins and constitute a crucial part of the interactions: the social constitution of water basins relies on the seamless interaction and transition between land and sea activities in a way that makes analytically separating them problematic. This shift of perspective can be very fruitful for IR, as it gives rise to entirely new spatialities and domains that go beyond landed units of analysis (see e.g. Phillips and Sharman, 2015). At the same time, the potential is not only one of exploring alternative geographies, but this has important implications for the theorisation of core IR concepts. The notion of littoral societies, for instance, puts forward that claim that within certain historically specific interaction circuits, 'the shore folk have more in common with other shore folk thousands of kilometres away on some other shore of the ocean than they do with those in their immediate hinterland' (Pearson, 2006, 353–4). Thus, a focus on medieval Monsoon Asia reveals distinct shared Islamic practices that are different from established orthodoxy and yet shared by a variety of merchant communities across a large littoral region (Prange, 2018). From an IR perspective, in which states and land-defined polities have traditionally been defined at cultural units, this opens up alternative ways of thinking of power, culture, and its circulation.

And yet, the notion of terraqueous itself already signals the insufficiency of bringing back the sea to IR through a purely anthropocentric lens. Although interest in ecologies and the environment has dramatically increased in IR in recent years, this has centred around the urgent political questions of the politics of climate change and the Anthropocene. While certainly relevant for the topic of IR's engagement with the sea, however, these perspectives are limited in their scope and attention to a specific – if crucial – aspect of contemporary international politics. A terraqueous analytics offers in this sense a broader and complementary perspective from which to tackle IR's entanglement with the sea. As Bashford notes, the lineage of the word 'terraqueous' points to an 'expansive Early Modern meaning, signaling not just land and sea, but vapours, airs, and waters' (Bashford, 2017: 255). In doing so, it calls attention to fundamental ecological and environmental ways of understanding water basins, as well as to specific ecologies and their role in human society. At a very basic level, thus, the question arises for IR scholars of how to analyse the relevance of different hydrological formations: can lacustrine, riverine, or oceanic relations, with their distinct ecological systems, be easily subsumed under the banner of 'the sea in International Relations'? In what way do the specific hydrological, ecological, and geographic conditions that constitute different sea-water basins interact with and give rise to social, political, and economic dynamics?

Environmental historians, following in the steps of Braudel, have long tackled precisely this independently of – but still connected to – current concerns about climate change. For indeed this perspective shows that ecological dimensions of water basins are fundamental to understanding the evolution of human societies even beyond human beings' abilities to shape their environment. In *The Corrupting Sea*, for example, Horden and Purcell unpack how the extreme ecological diversity in the Mediterranean, and the variability and uncertainty of these diverse microregions, interact to develop specific cultural systems that are oriented towards exchange and interaction (Horden and Purcell, 2000). The Mediterranean in this view would thus not just be a matter of happenstance evolution of social practices, but is closely related to a set of complex ecological process on which it in turn acts. From an IR perspective, Goettlich's chapter in this volume shows how the 'natural' world – in this case the oceans and fish – has had a fundamental impact in the evolution of core IR categories, such as that of territory. Similar types of analysis about the close imbrication between localised human and ecological processes in ways that configure individual water basins, regions, and ultimately spatialities can be conducted at various levels in IR.

Conclusion

As a concept with roots in an early modern expansive understanding of the global that sought to encompass both processes and matters, international terraqueous relations not only offers a valuable way of bringing the sea into IR. It also offers a vantage point from which to see why such an endeavour is crucial if IR wants to gain a better understanding of its subject matter. The invitation that this volume puts forward is a reminder that in doing so, IR should not limit itself to folding the sea back into a field that has a habit of stream-lining complexity in clear-cut dichotomies: in this case, first and foremost, the 'land' and the 'sea'.

In this conclusion we have argued that adopting a terraqueous perspective can not only help overcome these dichotomies but also put forward a distinct analytical perspective that contributes to a better understanding of international dynamics. As a set of questions, international terraqueous relations is a powerful counter to the Western *thalassodicy* that has underpinned the land/sea binary. In foregrounding an analytics of connection and discon-nection, it not only challenges teleological accounts of global-isation and of international encounters, but it also prevents the substantivist reification of entities that is imbricated in the repro-duction of global hierarchies. Finally, by drawing attention to the ways in which land and sea are constantly entangled in everyday processes, it can help us gain a better understanding of the myriad forms and nature of international terraqueous relations, past and present.

Note

1 Even naval powers like Spain and Portugal are not able to make the choice to become sea powers because their attachment to 'Monarchical absolutism, the Roman Church, terrestrial ambition, aristocratic priv-ilege, monopolistic economic models and a sustained contempt for seafarers, oceans and new ideas ensured they remained profoundly con-tinental' (Lambert, 2018: 204–5).

References

Armitage, D. (2002). 'Three concepts of Atlantic history', in D. Armitage and M. J. Braddick, eds, *The British Atlantic World, 1500–1800*, Basingstoke: Palgrave Macmillan, 11–27.

Armitage, D. (2018). 'The Atlantic Ocean', in A. Bashford, D. Armitage and S. Sivasundaram, eds, *Oceanic Histories*, Cambridge: Cambridge University Press, 85–110.

Barkawi, T., and M. Laffey (2006). 'The postcolonial moment in security studies', *Review of International Studies* 32(2): 329–52.

Bashford, A. (2017). 'Terraqueous histories', *The Historical Journal* 60(2): 253–72.

Baucom, I. (2001). 'Introduction: Atlantic genealogies', *The South Atlantic Quarterly* 100(1): 1–13.

Bell, D. (2020). *Dreamworlds of Race: Empire and the Utopian Destiny of Anglo-America*, Princeton, NJ: Princeton University Press.

Bertrand, R. (2011). *L'histoire à parts égales. Récits d'une rencontre Orient-Occident (XVIᵉ–XVIIᵉ siècle)*, Paris: Editions du Seuil.

Braudel, F. (2017 [1979]). *La Méditerranée et le monde méditerranéen à l'époque de Philippe II*, 4th rev. edn, Paris: Armand Colin.

Buzan, B., and G. Lawson (2014). 'Rethinking benchmark dates in international relations', *European Journal of International Relations* 20(2): 437–62.

Çapan, Z. G. (2020). 'Beyond visible entanglements: connected histories of the international', *International Studies Review* 22(2): 289–306.

Costa López, J. (Forthcoming). 'The premodern world', in M. Bukovanski, E. Keene, M. Spanu and C. Reus-Smit, eds, *The Oxford Handbook of History and International Relations*, Oxford: Oxford University Press.

Dudden, A. (2018). 'The sea of Japan/Korea's East Sea', in A. Bashford, D. Armitage and S. Sivasundaram, eds, *Oceanic Histories*, Cambridge, Cambridge University Press, 182–208.

Enloe, C. (2000 [1989]). *Bananas, Beaches and Bases: Making Feminist Sense of International Politics*, Berkeley, CA: University of California Press.

Fletcher, M. E. (1958). 'The Suez Canal and world shipping, 1869–1914', *The Journal of Economic History* 18(4): 556–73.

Getachew, A. (2019). *Worldmaking after Empire: The Rise and Fall of Self-determination*, Princeton, NJ: Princeton University Press.

Gilroy, P. (1993). *The Black Atlantic. Modernity and Double Consciousness*, Cambridge, MA: Harvard University Press.

Guillaume, X. (2011). 'The international as an everyday practice', *International Political Sociology* 4(5): 446.

Guillaume, X., and K. Grayson (2021). 'Sound matters: how sonic formations shape the nuclear deterrence and non-proliferation regimes', *International Political Sociology* 15(2): 153–71.

Guillaume, X., and J. Huysmans (2019). 'The concept of "the everyday": ephemeral politics and the abundance of life', *Cooperation and Conflict* 54(2): 278–96.

Hasty, W., and K. Peters (2012). 'The ship in geography and the geographies of ships', *Geography Compass* 6(11): 660–76.

Hedrick, A. (2015). 'Submarine sonar technicians detect, track, classify', *Navy News Service*, 18 March 2015, www.navy.mil/submit/display.asp?story_id=86108 (accessed 10 July 2020).

Horden, P., and N. Purcell (2000). *The Corrupting Sea: A Study of Mediterranean History*, Oxford: Blackwell.

Lambert, A. (2018). *Seapower States. Maritime Culture, Continental Empires and the Conflict That Made the Modern World*, New Haven, CT: Yale University Press.

Lambert, D., L. Martins and M. Ogborn (2006). 'Currents, visions and voyages: historical geographies of the sea', *Journal of Historical Geography* 32(3): 479–93.

Law, J., and A. Mol (2001). 'Situating technoscience: an inquiry into spatialities', *Environment and Planning D: Society and Space* 19(5): 609–21.

Lewis, M. W. (1999). 'Dividing the ocean sea', *The Geographical Review* 89(2): 188–214.

Linebaugh, P., and M. Rediker (2000). *The Many-headed Hydra: Sailors, Slaves, Commoners, and the Hidden History of the Revolutionary Atlantic*, Boston, MA: Beacon Press.

Ó Tuathail, G. (1996). *Critical Geopolitics: The Politics of Writing Global Space*, London: Routledge.

Osterhammel, J. (2014). *The Transformation of the World: A Global History of the Nineteenth Century*, Princeton, NJ: Princeton University Press.

Pearson, M. N. (2006). 'Littoral society: the concept and the problems', *Journal of World History* 17(4): 353–73.

Phillips, A., and J. C. Sharman (2015). *International Order in Diversity: War, Trade and Rule in the Indian Ocean*, Cambridge: Cambridge University Press.

Piot, C. (2001). 'Atlantic aporias: Africa and Gilroy's Black Atlantic', *The South Atlantic Quarterly* 100(1): 155–70.

Prange, S. R. (2018). *Monsoon Islam: Trade and Faith on the Medieval Malabar Coast, Cambridge Oceanic Histories*, Cambridge: Cambridge University Press.

Shilliam, R. (2015). *The Black Pacific: Anti-colonial Struggles and Oceanic Connections*, London: Bloomsbury.

Sivasundaram, S., A. Bashford and D. Armitage (2018). 'Introduction', in A. Bashford, D. Armitage and S. Sivasundaram, eds, *Oceanic Histories*, Cambridge: Cambridge University Press, 1–28.

Steinberg, P., and K. Peters (2015). 'Wet ontologies, fluid spaces: giving depth to volume through oceanic thinking', *Environment and Planning D: Society and Space* 33(2): 247–64.

Subrahmanyam, S. (1997). 'Connected histories: notes towards a reconfiguration of early modern Eurasia', *Modern Asian Studies* 31(3): 735–62.

Sullivan, A. K. (2010). 'Piracy in the Horn of Africa and its effects on the global supply chain', *Journal of Transportation Security* 3(4): 231–43.

Index

Ingram Content Group UK Ltd.
Milton Keynes UK
UKHW021941290623
424302UK00004B/29

9 781526 155108